Ralf Bruno Moura Lopes

Who's the boss here?

Ralf Bruno Moura Lopes

Who's the boss here?

Biotechnological advances and genomic perspectives for the use of Bacillus thuringiensis in agriculture

ScienciaScripts

Imprint
Any brand names and product names mentioned in this book are subject to trademark, brand or patent protection and are trademarks or registered trademarks of their respective holders. The use of brand names, product names, common names, trade names, product descriptions etc. even without a particular marking in this work is in no way to be construed to mean that such names may be regarded as unrestricted in respect of trademark and brand protection legislation and could thus be used by anyone.

Cover image: www.ingimage.com

This book is a translation from the original published under ISBN 978-3-330-75976-3.

Publisher:
Sciencia Scripts
is a trademark of
Dodo Books Indian Ocean Ltd. and OmniScriptum S.R.L publishing group

120 High Road, East Finchley, London, N2 9ED, United Kingdom
Str. Armeneasca 28/1, office 1, Chisinau MD-2012, Republic of Moldova, Europe
Printed at: see last page
ISBN: 978-620-8-31089-9

CONTENTS

"*There are no* victories without struggles, much less conquests without battles." Leonardo Luis Gomes

ACKNOWLEDGMENTS

To the Federal University of Viçosa, for the structure and training during my undergraduate and master's degrees.

To the Postgraduate Program in Agricultural Microbiology, for the training and the opportunity to carry out this work.

To the National Council for Scientific and Technological Development (CNPq) and the Coordination for the Improvement of Higher Education Personnel (CAPES) for their financial support.

To Professor Marisa Vieira de Q ueiroz, for her guidance, teachings and support in the research work.

To Professor Jeronimo Conceiçâo Ruiz of the René Rachou Research Center (CPqRR), for his help in annotating the genome.

To Professors Hilàrio Cuquetto Mantovani, Maria Cristina Dantas Vanetti and Juliana Lopes Rangel Fietto, for their advice and suggestions.

To the other professors in the Microbiology Department who contributed to my education.

To Lara, Elvira and Grace from the CPqRR for their help with note-taking and for their timely conversations.

To the staff of the Department of Microbiology, for their constant help.

To my parents, Ademilson and Rosângela, for their love, life lessons, support, dedication and effort that have allowed me to get this far.

To my godfather, José Tarcisio (*in memoriam*), and my aunts Lurdes (*in memoriam*) and Lili, for their love, encouragement and constant support.

To all my family members, for their support, friendship and trust.

To my friend Tiago Leite, for so many useful conversations about research and other subjects.

To all my friends at the Laboratory of Molecular Genetics of Microorganisms, for their friendship.

To all the people who, directly or indirectly, helped make this work possible.

1 INTRODUCTION

Bacillus thuringiensis is an endospore-forming Gram-positive bacterium. The species is widely distributed, being found in a variety of environments, such as soil, insect interiors, leaf surfaces and internal plant tissues. Most studies on *B. thuringiensis* have focused on direct effects against agriculturally relevant insect pests, although several new functions in plant protection against phytopathogens have been discovered. For example, the synthesis of zuitermicin A, an antibiotic that is mainly effective against oomycetes and related phytopathogenic fungi. Another example is the production of N-acyl homoserine lactonase, which hydrolyzes N-acyl homoserine lactone (AHL), a signal molecule of the bacterial *quorum sensing* system, suppressing the virulence of pathogens.

Endophytic bacteria colonize the interior of the host plant without causing apparent damage. Successful endophytic colonization by *B. thuringiensis* has been reported in many plants, such as beans, corn, sugarcane, cotton, soybeans, cabbage and lentils. Costa et al. (2012) reported that 32.9% of the isolates of cultivable endophytic bacteria from the leaves of different cultivars of the common bean (*Phaseolus vulgaris*) belonged to the Firmicutes phylum, including different isolates of *B. thuringiensis*. However, much remains unknown about the association of *B. thuringiensis* with legumes. Studies on this association have shown that *B. thuringiensis* inoculated in the rhizosphere or in seeds not only colonizes the interior of legumes, but also increases root nodulation and plant growth.

On the other hand, microbial diseases of the common bean are one of the main causes of the reduction in productivity of this important crop. The control of diseases caused by phytopathogens is usually done through a combination of crop management, pesticides and resistant cultivars. The application of pesticides is the main method of control, but their use is controversial due to their harmful effects on the environment and the selection of resistant pathogens.

There are also microbial diseases for which there is still no product with proven efficacy. Thus, the use of microorganisms capable of protecting the plant against phytopathogens has emerged as a more rational and safer alternative for crop management. In this context, endophytic bacteria can be useful for biological control of diseases, since they often occupy the same habitat as phytopathogenic microorganisms and cause no apparent damage to the host plant.

B. thuringiensis BAC3151 is an endophytic isolate of *P. vulgaris* that has demonstrated *in vitro* antimicrobial activity against phytopathogenic bacteria of the bean plant as well as potential for inhibiting *quorum sensing* in Gram-negative bacteria. In fact, genomic studies have revealed the *Bacillus* genus as a surprising potential source of antimicrobials, such as polyketides, non-ribosomal peptides as well as unusual antimicrobials and other secondary metabolites. These studies also make it possible to investigate events that change the architecture of the genome, such as gene loss or duplication, rearrangements and the acquisition of genes by horizontal transfer. In the present work, the genome of BAC3151 was sequenced in order to better investigate the genetic determinants potentially involved in the synthesis of antimicrobials that could be useful for the biocontrol of phytopathogens, in addition to analyzing genomic differences between the isolate and other strains.

2. LITERATURE REVIEW

2.1. Endophytic bacteria

Endophytic bacteria have been known *for* more than 120 years (HARDOIM et al., 2008) and correspond to microorganisms, cultivable or not, that colonize the interior of a host plant and cause no apparent damage to it (HALLMANN et al., 1997). Although nitrogen-fixing bacteria in the nodules of the common bean (*Phaseolus. vulgaris*) have been well studied and characterized (MARTiNEZ- ROMERO, 2003; MARTiNEZ-ROMERO, 2009) and the endophytic bacterial community of the seeds and roots of this legume has been reported (LÓPEZ- LÓPEZ, et al., 2010), little is still known about the endophytic bacteria of the aerial tissues of *P. vulgaris*. Costa et al. (2012) described the cultivable endophytic bacteria from the leaves of the Vermelhinho, Talismâ and Ouro Negro cultivars of *P. vulgaris* and found that 36.7% of the isolates belonged to Proteobacteria, 32.9% to Firmicutes, 29.7% to Actinobacteria and 0.6% to Bacteroidetes. The results also showed differences in the structure of the endophytic bacterial community according to bean cultivar.

The entry of microorganisms into plant tissue can generally occur through stomata, lenticels, lesions and the emerging surfaces of lateral roots and germinating rootlets (HUANG, 1986). However, the main entry points for endophytic bacteria appear to be naturally occurring lesions as a result of plant growth, in addition to root hairs and epidermal junctions (SPRENT and FARIA, 1988). Some bacteria also actively penetrate host tissues, producing cellulolytic and pectinolytic enzymes (HUREK et al., 1994).

Once inside the plant tissue, they can remain in a specific location, such as the root cortex, or colonize the plant systemically by means of transport through the conductive elements or the apoplast. The ability to disperse systemically has been demonstrated by endophytes such as *Erwinia* sp. in cotton (MISAGHI and DONNDELINGER, 1990), *Pseudomonas aureofaciens* in maize (LAMB et al., 1996) and *Bacillus thuringiensis* in legumes (TANUJA et al., 2013). The colonization of specific tissues by bacteria appears to be species specific. Generally, endophytic bacteria colonize the intercellular space (QUADT- HALLMANN and KLOEPPER, 1996), with few reports demonstrating intracellular colonization (FROMMEL et al., 1991; HUREK et al., 1994; QUADT- HALLMANN and KLOEPPER, 1996). Colonization of the vascular system has also been reported (GARDNER et al., 1982; LAMB et al., 1996).

The colonization patterns of endophytic bacteria in plant tissues are strongly dependent on biotic and abiotic factors that interact with each other (HALLMANN et al., 1997). One of the most important biotic factors is the presence of other microorganisms associated with the plant, such as fungi, viruses and other bacteria (HALMANN et al., 1998). As the space and nutrients provided by the host are limiting factors, there are various types of interaction between micro-organisms that have the same habitat, including antagonism, symbiosis and mutualism. The internal tissues of plants provide a more uniform and protective environment for micro-organisms than the plant surface, where exposure to extreme environmental conditions such as temperature, osmotic potential and ultraviolet radiation are limiting factors for survival (HALLMANN et al., 1997). An example of the influence of abiotic factors on the population of endophytic bacteria is what happens on sites contaminated with heavy metals, oil, solvents, polycyclic aromatic hydrocarbons or other organic contaminants, where plants growing in these soils are able to recruit microorganisms that can degrade the pollutants. Soil characteristics such as pH, salinity and texture also affect endophytic bacteria indirectly, as they alter the bacterial community of the rhizosphere, a source of potential endophytic bacteria (DOTY, 2008; RYAN et al., 2008).

In terms of their life strategy, endophytic bacteria can be classified as obligate or facultative endophytes. Obligate endophytic bacteria are strictly dependent on the host plant for their growth and survival. These bacteria are transmitted vertically or via a vector. Facultative endophytic microorganisms, on the other hand, can be characterized by alternating between the interior of plant tissues and the environment. The endophytic microbial diversity found in plants can be explained mainly by the ability of various facultative endophytes to enter and persist inside the plant (ROSENBLUETH and MARTINEZ-ROMERO, 2006).

In addition, endophytic bacteria can provide many advantages for plants, including growth promotion (BARKA et al., 2002; TANUJA et al., 2013), induction of systemic defense mechanisms (MISHRA et al., 2006; BAKKER et al., 2007), reduction of disease symptoms caused by various pathogens (COOMBS et al., 2004; KLOPPER et al., 2004), protection against insects and nematodes (D'ALESSANDRO et al., 2014), production of anti-herbivory compounds (SCOTT, 2001; SULLIVAN et al., 2007), biological nitrogen fixation (MARTiNEZ et al., 2003; JHA and KUMAR, 2007) and increased tolerance to environmental stresses (ANDREOLLI et al., 2013). Figure 1 shows some of the main types of interactions between endophytic bacteria and plants, involving the synthesis of metabolites with different roles.

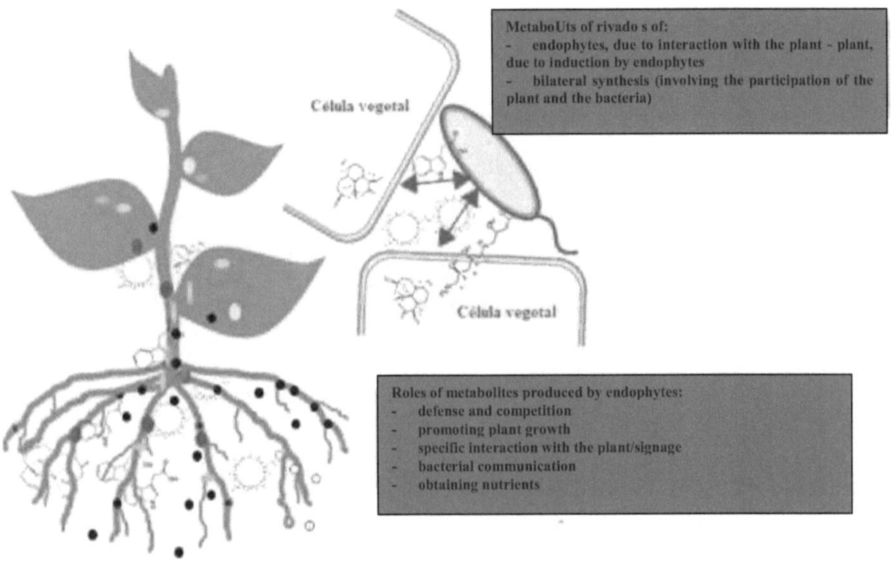

Figure 1. Schematic representation of different types of metabolites produced as a result of the interactions between endophytic bacteria and plants. The metabolites produced by endophytic bacteria can have important functions for the plant (BRADER et al., 2014).

1.2. *Bacillus thuringiensis*

1.2.1. General characteristics

Bacillus thuringiensis was first isolated from dead larvae of the silkworm (*Bombyx mori*) by bacteriologist Shigetane Ishiwata in Japan in 1901, while he was studying silkworm disease in this insect, and the species was initially named *Bacillus sotto* (ISHIWATA, 1901). However, in 1911, microbiologist Ernst Berliner isolated a related strain in larvae of the flour moth (*Anagasta kuehniella*) in the German state of *Thuringia* and subsequently named it *B. thuringiensis*, a name that has been retained (BERLINER, 1911; BERLINER, 1915).

B. thuringiensis is a Gram-positive, facultative anaerobic, mesophilic and chemoheterotrophic bacterium, growing mainly in the 10 to 40 °C range. It has the shape of a rod with dimensions of around 1.1 μm wide and 4.0 μm long, and is able to move around due to the presence of peritrichial flagella (depending on the serovar) (STAHLY et al., 1991). Under unfavorable environmental conditions, cell division is interrupted and the sporulation cycle typical of bacilli begins, generating an ellipsoidal spore located in the central or paracentral region of the cell. Simultaneously with the sporulation process, crystalline inclusions are produced, which are characteristic of each *B. thuringiensis* variety and are toxic to insects (FEITELSON et al., 1994; IBRAHIM et al., 2010). These crystalline inclusions, also called Cry proteins or δ-endotoxin, are released with the spore at the moment of cell lysis (HANNAY and FITZ-JAMES, 1995). In addition, they are an important characteristic for differentiating *B. thuringiensis* from similar species (FEDERICI, 1999).

1.2.2. Distribution in the environment

The *B. thuringiensis* species *is* widely distributed and can be found in almost all parts of the world and associated with various environments, such as soil, water, stored grain (BRAVO et al., 1998), dead and living insects (SUZUKI et al., 2004) and the surface and interior of plants (BRAVO et al., 1998; MONNERAT et al., 2003; COSTA et al., 2012).

However, there is no certainty about the primary habitat of *B. thuringiensis* (MEADOWS, 1993). One possibility is that the bacterium is originally an entomopathogen, since it would be unlikely to use a large amount of energy during sporulation to form a crystal that did not have great adaptive value. As the spore and the crystal released by the cell tend to remain together in the environment, after ingestion by a susceptible larva and the subsequent death of the insect, germination of the spore can then occur (THOMAS et al., 2000; VILAS-BÔAS et al., 2000 and SUZUKI et al., 2004).

A second hypothesis is that *B. thuringiensis* is native to the soil. The germination of spores in sterilized soil is described in several studies and depends on humidity, nutrient availability, pH, as well as the type of soil itself (AKIBA, 1986). However, later studies have shown that *B. thuringiensis* spores can often fail to germinate in non-sterile soil (VILAS- BÔAS et al., 2000) and that, in this type of soil, the death of inoculated vegetative cells is significantly higher than in sterile soil (THOMAS et al., 2000; FERREIRA et al., 2003). According to Arantes et al. (2002), this lower viability may be due to the low capacity of *B. thuringiensis* to multiply in this environment, possibly because it is not an original member of the soil microbial communities, remaining mainly in the form of spores.

Another less discussed possibility, but presented by Smith and Couche (1991), is that the species is native to the phylloplane of trees. The authors isolated *B. thuringiensis* from various tree species and in quantities high enough to be the result of dissemination by wind or rain alone. According to the authors, the soil is only a reservoir, while the primary habitat could be the phylloplane of trees.

1.2.3. Use for biocontrol in agriculture

B. thuringiensis is a species that produces a variety of compounds released into the extracellular medium that are potentially useful for different types of biocontrol in agriculture. These compounds include antibiotics (FICKERS, 2012), chitinases (GOMAA, 2012) and other antifungals (SADFI et al., 2001), bacteriocins (CHERIF et al., 2003), lactonases (DONG et al., 2004), siderophores (BODE, 2009) and the characteristic δ-endotoxin (Cry proteins), which is the most studied and widely used product of *B. thuringiensis* in the world (POLANCZYK and ALVES, 2003).

The first study using *B. thuringiensis* for biocontrol in the field was conducted by Husz (1929), as part of an international program to control the European corn caterpillar (*Ostrinia nubialis*). Although the cause of the bacterium's entomopathogenic activity was not yet known at the time, the results obtained were promising and from then on interest in the species grew. The first formulation based on *B. thuringiensis*, *Sporeine*, was produced in France in 1938. In the 1950s, large-scale production began with the launch of the commercial *bioinsecticide Thuricide*, followed by similar products from companies located in Russia, Germany, the United States and the former Czechoslovakia (DE MAAGD et al., 1999). Initially, the products were only used to control lepidopteran larvae. However, especially since the 1970s, new subspecies of the bacterium have proved effective against insects of the orders Diptera, Coleoptera (WEISER, 1986; EDWARDS et al., 1988), Hymenoptera, Hemiptera and Orthoptera (FEILTEISON, 1994). Toxicity has also been reported for species of nematodes, protozoa, mites and molluscs (FEILTEISON, 1994; WANG et al., 2013; LIU et al., 2014).

The efficacy and specificity of *B. thuringiensis* strains and their toxins in controlling insects favors the formulation of biopesticides based on this bacterium and more than one hundred formulations have been placed on the world market since the first product was launched, accounting for more than 90% of total sales of bioinsecticides (POLANCZYK and ALVES, 2003). The direct application of *B. thuringiensis* crystals and spores is the main way of using these biopesticides, which are harmless to the environment, beneficial insects and other animals, including humans (CARDENAS et al., 2001; BRAR et al., 2006).

In North America, products based on *B. thuringiensis* are widely used to control forest pests, mainly *Lymantria dispar, Choristoneura fumiferana* and *Choristoneura occidentalis*, as well as insects harmful to agricultural crops. In Australia, these bioinsecticides are used against pests of crops such as cotton, fruit plants, ornamental plants and tobacco (GLARE and O'CALLAGHAM, 2000; VAN FRANKENHUYZEN, 2000). They are also used against insects that damage large crops, forests and vegetables in approximately one million hectares covering around thirty provinces in south-central China. In Egypt, *B. thuringiensis* is mainly used to control cotton pests (SALAMA and MORRIS, 1993). Cuba and Mexico lead the way in the use of bioinsecticides based on this microorganism in Latin America, especially for insect management in cotton, banana, potato, citrus, vegetables, tobacco, corn and pasture crops. In Brazil, around thirty agriculturally important pests and mosquito vectors are controlled by *B. thuringiensis* (POLANCZYK and ALVES, 2003).

Despite their widespread use, the cost of bioinsecticides based on *B. thuringiensis* is often higher than that of chemical insecticides, the crystals cannot penetrate plants and the persistence of most formulations in the field is low (GELERNTER and SCHWAB, 1993; VAN FRANKENHUYSEN, 1993; NAVON 2000). In this sense, research to minimize these limitations is being carried out, mainly through the prospecting of new *cry* genes, the production of plants expressing these genes, the development of new formulations and also the expansion of knowledge of the interactions of *B. thuringiensis* with other entomopathogens and with chemical insecticides (NAVON, 2000).

1.3. A brief history of genomic science

With the advent of DNA sequencing technology based on didesoxynucleotides, developed by Sanger et al. (1977b), the first genome sequencing projects were made possible. In the same year, this methodology enabled the complete DNA sequencing of the ΦX174 bacteriophage (SANGER et al., 1977a). Despite this initial advance, it was only after years of technological development using the methodology that the first genomes of cellular organisms, the bacteria *Haemophilus influenzae* and *Mycoplasma genitalium*, were completely sequenced (FLEISCHMANN et al., 1995; FRASER et al., 1995). From then on, genomic science began to develop in conjunction with bioinformatics, completely changing the biological landscape and characterizing the most diverse organisms on the tree of life.

Mainly due to the smaller size of their genomes, the *Bacteria* and *Archaea* domains have the largest number of organisms with finalized genomes and ongoing genomic projects. There are currently a total of 39,183 bacterial genome projects, with Proteobacteria, Firmicutes and Actinobacteria accounting for 43.8%, 30.8% and 14.5%, respectively, of these projects (https://gold.jgi-psf.org/distribution) (data up to

December 2014) (Figure 2). The predominance of these phyla is mainly explained by the priority given to sequencing the genomes of organisms that have already been studied and characterized by research groups around the world (KYRPIDES, 2009).

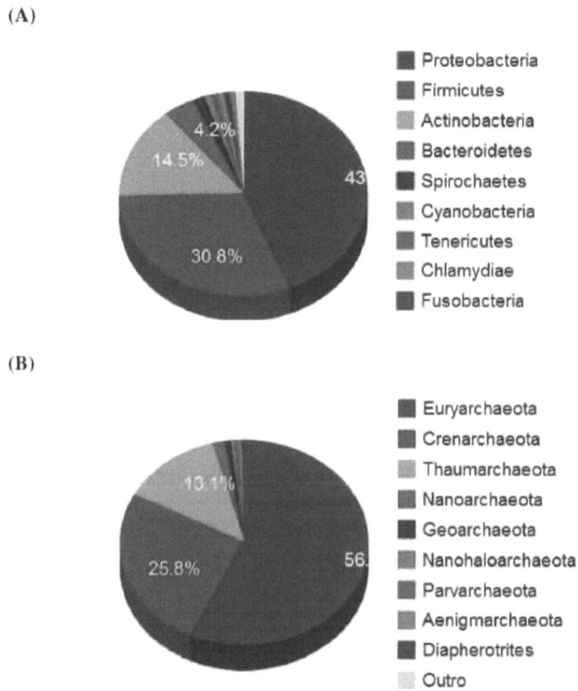

Figure 2. Phylogenetic distribution of (A) *Bacteria and* (B) *Archaea* genome projects registered in the *Genomes Online Database* (GOLD) (https://gold.jgi-psf.org/distribution) up to December 2014.

Since the publication of the *draft* human genome (VENTER et al., 2001), the ability to generate genomic data at lower cost, faster and more efficiently has increased even more, especially due to the development of new DNA sequencing technologies, with the so-called next-generation sequencers. There are several technologies available on the market for large-scale DNA sequencing. These technologies differ mainly in terms of the mechanism used in the sequencing process, the volume of data generated, the accuracy of bases, the size of the fragments generated and the operating cost (MACLEAN et al., 2009).

In the first genome sequencing projects using Sanger technology, the fragments sequenced were small in number and up to 1000 bp in size, and were processed over a period of weeks or months in exhausting laboratory work. However, at the beginning of this millennium, the automation of sequencing technologies was perfected to the point where fully automated equipment began to generate sequences in parallel 24 hours a day and large genome sequencing centers around the world housed dozens of these sequencing machines. This, in turn, led to the need to create new genome assembly algorithms, using sequences of generally 35 to 1000 bases in length, with sequencing error rates ranging from 0.5 to 15%, and which can still contain complex artifacts such as repeats (MARDIS, 2008).

The growing advances in technology have also led to the emergence of various consortia and large-scale projects around the world. The Human Microbiome Project was initiated with the aim of characterizing the entire microbial community associated with the different tissues of the human organism and the expected results could contribute to a better understanding of human physiology and predisposition to disease (TURNBAUGH et al., 2007). The *TerraGenome* project was set up to sequence the complete metagenome of different soil samples from around the world (http://www.terragenome.org/). Finally, the largest and most ambitious current genome sequencing project, the *Earth Microbiome Project*, which began in 2010, aims to characterize 200,000 samples from different environments around the world. Using metagenomic, metatranscriptomic and amplicon sequencing approaches, the researchers plan to reconstruct approximately 500,000 microbial genomes and build a gene atlas that will contain information and metabolic models specific to each biome sampled (http://www.earthmicrobiome.org/).

1.4. Genome assembly

The assembly of a genome consists of a set of procedures in which a large number of short DNA sequences are organized in a linear space with the aim of representing the complete DNA molecule(s) that make up the genome of the species studied (DOHM et al., 2007).

The huge amount of DNA sequence data generated by next-generation sequencing

(NGS) technologies, together with the artifacts and errors inherent in each sequencing technology, challenge projects to assemble complete genomes from different species. Less than a decade ago, only the *Newbler* assembler was available for this purpose. It was designed to assemble fragments generated by sequences produced by the Roche 454 sequencer (proprietary *software*). Presented in mid-2007, the hybrid version of the MIRA assembler (CHEVREUX et al., 2004) was the first free-use assembler developed to assemble segments from the 454 sequencer and mixtures of segments from the 454 sequencer and the Sanger method. At the end of 2007, the SHARCGS assembler (DOHM et al., 2007) was developed to assemble short segments from Illumina technology and was quickly followed by a series of other *software packages*. Currently, there are several options for assemblers capable of using different fragment sizes, numerous file formats and which are applied to genomes of different complexities (HUNT et al., 2014).

Genome assembly algorithms work by taking all the DNA segments at once, aligning them with each other and trying to identify the regions where two read segments overlap. These overlaps can be incorporated linearly into the assembly process. The shorter the sequences, the greater the number of overlaps required to perform the assembly. In addition, genomic coverage, i.e. the number of times a given region of the genome is covered by read segments, contributes to increasing accuracy in the region considered (EKBLOM and WOLF, 2014).

The use of reading segments with paired ends, i.e. with overlapping DNA sequences at both ends, facilitates the process of obtaining assembled sequences. When two or more of the sequences obtained from the sequencing process are paired (because they have paired ends), larger sequences called *contigs* are generated. In turn, *contigs* can be paired, generating even larger sequences called *scaffolds*. Regions with no overlap are usually found, characterizing *gaps* (BAKER, 2012). Determining the sequence present in the *gaps* can be achieved by specifically amplifying these regions with subsequent resequencing, making it possible to achieve the final assembly of the genome (ALKAN et al., 2011).

NGS technologies easily generate two or three billion DNA read segments with 100 copies each, which can be used to assemble the species' genome (BAKER, 2012). Assembly is, of course, a highly complex challenge due to the small size of the initial sequences. During genome assembly, the read fragments are usually aligned with a genomic sequence known as the "reference" for the assembly (HUNT et al., 2014). In the absence of a reference genome, the read sequences must be used for *de novo* assembly. The term *de novo* comes from Latin and means "from the beginning". It therefore refers to the methods used to determine the complete DNA sequence when there is no genomic sequence available to use as a reference. The different *de novo* assembly strategies have advantages and disadvantages in terms of speed and accuracy when compared to each other. *De novo* assembly *is* almost always complex and difficult, particularly when the genome is large and has sequences that are repeated many times, causing the assembly to fail (ALKAN et al., 2011). The decision to use the *de novo* or reference-based assembly strategy, if the latter is available, is based on biological application, cost, effort to achieve the required accuracy and assembly time considerations (EKBLOM and WOLF, 2014).

New advances in the assembly process are expected with the integration of databases and the exploration of multiple sequencing strategies, always with the aim of meeting the challenge of assembling complex genomes. As mentioned, large genomes marked by an abundance of repeated sequences are still a major challenge for the development of assembly algorithms based on short DNA sequences. Usually, several assemblers are combined to get around this problem. Meanwhile, the accuracy and length of the sequenced fragments are gradually increasing (ZHANG et al., 2011). The challenge is to optimize the assembly algorithms to be able to process the vast amount of information from NGS sequencing, which has its own particularities for processing, both in terms of volume and the types of data specific to each sequencer (CHAISSON and PEVZNER, 2008).

1.5. Genomic annotation

1.5.1. Automatic note-taking

The interpretation of raw genomic data involves the identification and annotation of various elements present in DNA, such as protein coding sequences, tRNAs and rRNAs, pseudogenes and repetitive DNA (MEDIGUE and MOSZER, 2007). The scientific community's interest in genomic annotation includes clarifying the biological data obtained and placing it in a context for understanding biological processes (STEIN, 2001). Data can be interpreted at the nucleotide, amino acid and biological process levels.

When the annotation is at the nucleotide level, the main aim is to predict probable genes or *open reading frames* (ORFs) (LUKASHIN and BORODOVSKY, 1998; STEIN, 2001). The programs that perform this task generally use statistical calculations to recognize intrinsic characteristics of the coding regions (STEIN, 2001). The use of orthologous genes from other organisms is also an important strategy for gene prediction (STEIN, 2001; MÉDIGUE and MOSZER, 2007).

After predicting genes, it is important to infer which products they encode and in which biological processes they will act (STEIN, 2001). Various programs are used in these annotation steps, not only to identify products and their functions, but also to standardize nomenclatures in a universal way (REEVES et al., 2009). Automatic annotation provides a final map made up of several characterized genetic elements, the content of which will be improved in manual curation. Table 1 shows the main programs used in automatic annotation and their respective functions.

Table 1. Computational tools widely used to annotate bacterial genomes.

Use	Program	Reference
Gene prediction	FgenesB Glimmer	http://www.softberry.com/ Delcher et al., 1999
tRNA prediction	tRNA-Scan-SE ARAGORN	Lowe and Eddy, 1997 Laslett et al., 2004

rRNA prediction	RNAmmer Rfam	Lagesen et al., 2007 Griffiths-Jones et al., 2005
Prediction of repetitive elements	*Repeat Scout*	Price et al., 2005
Prediction of protein motifs, domains and families	InterProScan	Hunter et al., 2009
Signal peptide prediction	SignalP	Bendtsen et al., 2004
Predicting promoters	BPROM PromScan BDGP	http://www.softberry.com/ Studholme and Dixon, 2003 Reese, 2001
Prediction of pseudogenes	Consed	Gordon et al.,1998

1.5.2. Manual curation

Manual curation is a crucial stage for finalizing and validating the automatic annotation of a given genomic sequence (MÉDIGUE and MOSZER, 2007). It is at this stage that human evaluation (by the curator) becomes necessary in order to correct and legitimize the genomic information previously predicted (STEIN, 2001). As a result of manual curation, there is an increase in the accuracy of the data that will be made available to the scientific community (REEVES et al., 2009) and, consequently, a reduction in the subsequent propagation of errors arising from automatic annotation (GUNDOGDU et al., 2007).

Manual curation takes place through similarity search analyses in databases containing well-characterized and conserved proteins, thus providing a more detailed description of the product of a given gene or gene family (OVERBEEK et al., 2005). In addition, it is now possible to characterize the metabolic pathways, domains and conserved protein motifs present in the genome as well as to use a uniform name for enzymes and other products (RUST et al., 2002).

Some groups working on prokaryote genomics have made efforts to manually re-annotate genomes already deposited in databases (GUNDOGDU et al., 2007; LUO et al., 2009). These efforts are based on needs such as updating the information in various genomes that is out of date in the databases, providing more detailed descriptions for their genes and respective proteins (GUNDOGDU et al., 2007), as well as correcting or reducing errors arising from automatic annotation, which can be incomplete or lead to false descriptions (MADUPU et al., 2010), thus making the available genomic information more accurate.

It is also known that databases contain around 30% to 40% of proteins with no known function or hypothetical proteins (HANSON et al., 2010). A flaw in these databases is that in many of them, such as *GenBank*, the information is not updated periodically, becoming obsolete and leading to errors in new annotations (SALZBERG, 2007). Many of the proteins annotated as hypothetical already have a description available today, as well as conserved domains that can help describe their functions and their updating has been carried out in the process of manual curation, both unpublished and linked to re-annotation (GUNDOGDU et al., 2007; LUO et al., 2009).

In fact, other approaches make use of the results obtained post-curation, such as comparative genomics approaches. Genomic comparison between related species or between strains of a given microorganism can then be carried out with greater certainty (RUST et al., 2002).

1.6. *B. thuringiensis* genome projects

There are currently 42 *B. thuringiensis* genomes deposited at the *National Center for Biotechnology Information* (NCBI) (http://www.ncbi.nlm.nih.gov/ genome/browse/), of which 15 are closed genomes and 27 are *draft* genomes (genomes not completely assembled) (data up to December 2014).

The genomes deposited for the species range from 4.93 Mb (for *B. thuringiensis* JM-Mgvxx-63) to 7.09 Mb (for *B. thuringiensis* T01-328), with a total of 135 plasmid sequences identified and made available in the closed genomes. In the *draft* genomes, it is possible that some *contigs* or *scaffolds* may correspond to plasmids, leading to an increase in this total value with the closure of other genomes, although certain strains may not have plasmids. The strains with larger genomes also have a greater number of genes, with *B. thuringiensis* T01-328 having 7518 genes, while *B. thuringiensis* JM-Mgvxx-63 has 4931 genes. The number of protein-coding sequences for these two strains is 7403 and 4719, respectively. The G+C content between all the sequenced genomes does not vary much and remains between 34.5 and 35.5%.

Interest in the sequenced *B. thuringiensis* strains is mainly focused on their use in biocontrol, due to their characteristics such as insecticidal activity (GUAN et al., 2012; LIU et al., 2013; MURAWSKA et al., 2013; PALMA et al., 2014), nematicidal activity (LIU et al., 2014), molluscicidal activity (WANG et al., 2013) and chitinolytic activity (MARCON et al., 2014). Other investigations into some of the strains include their use for genetic engineering, due to their high transformation efficiency (HE et al., 2010; JEONG et al., 2014), the study of biofilm formation (ZWICK et al., 2012) or even their involvement in human pathology (a particular case) (HAN et al., 2006).

1.7. Genome mining and bioprospecting of *Bacillus* products

Recent advances in genome sequencing have highlighted the genus *Bacillus* as an unexpectedly promising source of bioactive compounds. In fact, for some of them, such as *Bacillus subtilis*, more than 4% of the genome has been found to be potentially involved in the synthesis of compounds such as non-ribosomal peptides, polyketides and unusual antimicrobials (KUNST et al., 1997; ARGUELLES-ARIAS et al., 2009; CHEN et al., 2009).

Non-ribosomal peptide synthetases (NRPSs) are large multimodular enzymes that synthesize non-ribosomal peptides from the successive binding of activated amino acids. In addition to the main domains that catalyze the specific reaction for the incorporation of an amino acid, NRPSs have more specialized domains that carry out modifications to these monomers, such as epimerization, methylation, oxidation, reduction, formylation and heterocyclization. The specialized domains allow NRPSs to synthesize a large number of diverse structures with a variety of biological functions that could not be achieved by the ribosomal machinery (FELNAGLE et al., 2008). Structural diversity is responsible for the different biological activities, targets and mechanisms of action of non-ribosomal peptides, making it possible to exploit these peptides biotechnologically (FICKERS, 2012).

Since the characterization of the first non-ribosomal peptide, gramicidin produced by *Bacillus brevis*, in the early 1970s, many others have been isolated and their biosynthetic *clusters* characterized. With the emergence of powerful molecular and bioinformatics tools, the identification of even new metabolites through genome mining has become a reality (ANSARI et al., 2004; VAN LANEN and SHEN, 2006). For example, studies of *B. thuringiensis* genomes have led to the prediction of a new NRPS consisting of seven modules that could be involved in the synthesis of a heptalipopeptide similar to kurstacin (ABDERRAHMANI et al., 2011). The discovery of new natural compounds using genome mining is an encouraging sign, suggesting that this methodology could lead to the isolation of new molecules of biotechnological interest.

In turn, polyketide synthases (PKSs) produce the central structure of polyketides from acyl-coenzyme A monomers (JENKE-KODAMA and DITTMANN, 2009). Type I PKSs are similar to NRPSs in that different catalytic domains are found in a single polypeptide chain, while type II PKSs present the catalytic domains in mono- or bifunctional proteins (HERTWECK et al., 2007). In addition to the central domains involved in biosynthesis, additional PKS domains such as ketoreductase, dehydratase, methyltrasferase, enoyl reductase and oxidase are responsible for the chemical modification of the growing polypeptide. Type II PKSs often have a cyclase domain responsible for the formation of aromatic structures (MEURER et al., 1997). Similar to non-ribosomal peptides, due to their versatile assembly mechanism, polyketides exhibit remarkable diversity in terms of both structure and biological activity. For example, the genes involved in the synthesis of dificidin, macrolatin and bacillin have been characterized in *Bacillus amyloliquefaciens* (CHEN et al., 2007; ARGUELLES-ARIAS et al., 2009). While dificidin has a broad spectrum of antibacterial activity (WILSON et al., 1987; ZIMMERMAN et al., 1987) and has been promising in antagonizing *Erwinia am \lovara*, a phytopathogen that causes fire blight in various rosaceae, such as apples and pears (ZWEERINK and EDISON, 1987), macrolatin and bacillene have a shorter spectrum and have not been associated with biocontrol in agriculture, but have potential for use in medicine (GUSTAFSON et al., 1989; PATEL et al., 1995).

In addition to non-ribosomal peptides and polyketides, several *Bacillus* can also produce interesting antimicrobial peptides synthesized by ribosomes, the bacteriocins. Strains *of B. thuringiensis*, for example, synthesize a variety of bacteriocins and bacteriocin-like inhibitory peptides. The antimicrobial peptides *of B. thuringiensis* are mainly turicins, tocicins and entomocins, such as turicin B439 from *B. thuringiensis* B439 (AHERN et al., 2003), turicin 17 from *B. thuringiensis* NEB17 (LEE et al., 2009b), turicin H from *B. thuringiensis* SF361 (LEE et al., 2009a), turicin S from *B. thuringiensis* sv. *entomocidus* HD198 (CHEHIMI et al., 2007), tocicin from *B. thuringiensis* sv. *tochigiensis* HD868 (PAIK et al., 1997), entomocin 9 from *B. thuringiensis* sv. *entomocidus* HD9 (CHERIF et al., 2003) and entomocin 110 from *B. thuringiensis* HD110 (CHERIF et al., 2008). *B. thuringiensis* sv. *tochigiensis* BGSC4Y1 and *B. thuringiensis* sv. *monterrey* BGSC4AJ1 can produce subtilosin (READ et al., 2009). Other antimicrobial peptides such as morricin 269, kurstacin 287, queniacin 404 and tolvorticin 524 from Mexican strains have also been reported (BARBOZA-CORONA et al., 2007; DE LA FUENTE-SALCIDO et al., 2008).

Examples of bacteriocins with potential for biocontrol in agriculture include ericin S, produced by *B. subtilis* A1/3, which is active against *Clavibacter michiganensis*, the causal agent of bacterial canker of tomato, and the peptide Bac 14, produced by *B. subtilis* 14B isolated from rhizosphere, active against *Agrobacterium tumefaciens* (ABRIOUEL et al., 2011). Other applications of bacteriocins in agriculture are also possible. The promotion of plant growth by turicin 17 produced by *B. thuringiensis* NEB17 has been demonstrated (LEE et al., 2009 a,b). Peptides with antifungal activity can also be applied in the control of plant decay and in the post-harvest control of fruit and vegetables, as in the case of the peptide produced by *B. amyloliquefaciens* RC-2, active against *Colletotrichum demantium* and other phytopathogenic fungi and bacteria (ABRIOUEL et al., 2011).

In Gram-negative bacteria, the population density-dependent gene expression control mechanism, *quorum sensing*, involves signaling molecules, especially N-acyl homoserine lactones (AHLs) (FUQUA et al., 2001), and phytopathogenic bacteria can use this mechanism to regulate the production of virulence factors (VON BODMAN et al., 2003). However, *quorum sensing* can be inhibited by enzymes that degrade AHL, such as lactonases. Dong et al. (2004) showed that *B. thuringiensis* strains producing N-acyl homoserine lactonase suppress the *quorum sensing-dependent* virulence of the bacterial pathogen *Pectobacterium carotovorum* (*Erwinia carotovora*). Alternatively, plants genetically modified to express *aiiA*, the N-acyl homoserine lactonase gene from *Bacillus*, have been studied. Overexpression of this gene in tobacco plants, for example, significantly reduced the incidence of wildfire caused by *Pseudomonas syringae* pv. *tabaci* (QUINONES et al., 2005). These results show the potential for exploiting antagonists that inhibit the bacterial cell communication mechanism for disease control and prevention.

In addition, plant growth and production can also be seriously affected by abiotic stresses. Among these stresses, iron deficiency is the main factor leading to reduced productivity, especially in calcareous soils, where iron solubility is extremely low (KOBAYASHI et al., 2005). One of the mechanisms most used by microorganisms to acquire iron is the production and excretion of siderophores, which are low molecular weight chelating agents that bind to Fe^{3+} with high affinity and transport it into the cell (ZAWADZKA et al., 2009). Thus, one way in which plant-associated micro-organisms that produce siderophores contribute to plant health is by complexing these compounds with iron, making it less available to phytopathogens (CHEN et al., 2009). It has also been shown that, under experimental conditions of sterile soil, plants show symptoms of iron deficiency and have very low levels of the element in the root, which highlights the importance of microorganisms for the acquisition of iron by the plant (RADDADI et al., 2007). For *Bacillus*, the production of various siderophores has been detected, mainly schizokinene from *Bacillus megaterium* and others from *B. subtilis*, *Bacillus licheniformis* and the *B. cereus* group. Bacilibactin and petrobactin, for example, are produced by *B. cereus* (PARK et al., 2005), *Bacillus anthracis* (CENDROWSKI et al., 2004) and *B. thuringiensis* (WILSON et al., 2006).

Another class of *Bacillus* products is chitinases. Chitin is an insoluble polysaccharide made up of linear chains of N-acetylglycosamine residues linked by β-1,4-glycosidic bonds. It is one of the most abundant biopolymers on the planet, second only to cellulose, and is an important component of the cell wall of fungi, as well as being found in the external skeleton of insects and also in algae, crustaceans and the internal structures of other invertebrates (BHATTACHRYA et al., 2007). Each year, a vast amount of chitin waste is released into the environment, which can cause environmental problems, and the decomposition of this polymer is important for maintaining the ecosystem's carbon and nitrogen balance (HAYES et al., 2008). Chitin catabolism involves the initial cleavage of the polymer by chitinases, producing oligosaccharides, and then subsequent hydrolysis into N-acetylglucosamine monomers by chitobiases (SUGINTA et al., 2000). These derivatives have a variety of agricultural, industrial and medical applications, including use in cosmetics and antitumor activity (BANSODE and BAJEKAL, 2006).

Several *Bacillus* species, such as *Bacillus alvei*, *Bacillus lentus*, *Bacillus licheniformis*, *Bacillus cereus* and *B. thuringiensis*, can produce chitinases during growth (ROJAS-AVELIZAPA et al., 1999; GOMAA, 2012). Chitinases are used as a potential biocontrol agent for many pathogenic fungi and can thus be useful for plant disease control (TANG et al., 2012), food and seed preservation (DE LA VEGA et al., 2006), as well as bioremediation (WANG and HUANG, 2001) and the preparation of oligosaccharides and N-acetyl-D-glucosamine (WANG et al., 2006). Additionally, the potential role of chitinases in the biocontrol of insects, such as mosquitoes, has been suggested (HAYES et al., 2008). Chitinases can also be used in the area of human health, such as in the composition of ophthalmic preparations, and in other biotechnological areas (DAHIYA et al., 2006).

Thus, many *Bacillus* products have been discovered in recent years and, although great progress has been made in their bioprospecting, new compounds of biotechnological interest are still expected to be found as well as the optimization of their production.

3. MATERIAL AND METHODS

3.1. Bacterial isolate and growth conditions

Bacillus thuringiensis BAC3151 belongs to the bacterial library of the Laboratory of Molecular Genetics of Microorganisms/BIOAGRO at the Federal University of Viçosa (UFV). The isolate was previously obtained from the leaves of *P. vulgaris* (COSTA et al., 2012) and cultivated in 10% TSA (*Tryptone* Soy *Agar*) medium (tryptone 1.5 g/l, soy peptone 0.5 g/l, NaCl 1.5 g/l, pH 7.3) at 28°C. The stock was kept at -80°C in cryotubes containing 10% TSA medium and 40% glycerol.

3.2. DNA extraction and genome sequencing

For genome sequencing, total DNA was isolated from the culture grown for 16 hours using the *Wizard genomic DNA purification kit* (Promega), according to the manufacturer's instructions for Gram-positive bacteria. The purified total DNA (25.278 μg and $A_{260}/A_{280} = 2.05$) was sequenced by Macrogen Inc. (Seoul, South Korea) using the HiSeq 2000 (100 bp *paired-end*) sequencing system (Illumina). A total of 21,512,378 high-quality *reads* and 2,172,750,178 high-quality bases were obtained, producing a coverage of approximately 355X.

3.3. Data processing and assembly

The raw sequencing data was processed using the Phred-Phrap-Consed package (RASKO et al., 2008). The minimum quality value used in the *base-calling* program was Q20. Assembly was then carried out using *SOAPdenovo* version 2.04 (LUO et al., 2012).

3.4. Genome annotation

The annotation process involved the use of several algorithms in a multi-step procedure. Structural annotation was carried out using the following *software*: FgenesB: gene predictor (http://www.softberry.org/); RNAmmer: rRNA predictor (LAGESEN et al., 2007) and tRNAscan-SE: tRNA predictor (LOWE and EDDY, 1997). Functional annotation was carried out by similarity analysis with BLAST (ALTSCHUL et al., 1990), using the non-redundant (NR) bank and the conserved domain (CDD) bank, and analysis with InterProScan (JONES et al., 2014) (*e-value* <1e-6). In addition, manual annotation was carried out with Artemis (RUTHERFORD et al., 2000).

3.5. Phylogenomic analysis

Phylogenomic analysis of *B. thuringiensis* was based on the *core* genes of BAC3151 and 27 other strains obtained from NCBI at ftp://ftp.ncbi.nih. gov/genomes/. In the process, the *core* genes were aligned with MUSCLE version 3.8.31 (EDGAR, 2004), the non-matching regions of the alignment were masked and subsequently removed by GBLOCKS (CASTRESANA, 2000). The remaining regions of all the alignments were then concatenated to form a large sequence alignment, which was then used as input for PHYLIP v. 3.696 (FELSENSTEIN, 1989). The tree was built using the Neighbor-Joining method with 1000 replicates.

3.6. Arrangement and orientation of the *scaffolds*

The order and orientation of the *B. thuringiensis* BAC3151 genome *scaffolds* was determined using the Mauve v. 2.3.1 *software* using the *Mauve Contig Mover* algorithm (DARLING et al., 2010), using the *B. thuringiensis* MC28 sequence as a reference and the default parameters.

3.7. Gene sharing between strains

Efficient Database framework for comparative Genome Analyses using BLAST score Ratios (EDGAR) (BLOM et al., 2009) was used to analyze the pangenome and *core* genome *of B. thuringiensis*, as well as the unique coding sequences (*singletons*) of BAC3151. In the process, the results of the reciprocal BLAST between groups of genomes are taken into account. Protein coding sequences (CDSs) from a genome that do not have a *score ratio value* (SRV) (*observed* score/maximum *score*) greater than the cut-off point in any of the other genomes analyzed are considered *singletons*. The SRV of the cut-off point (0.33) was generated during the creation of the bank of genomes used in the analysis. When calculating the *core* genome, only the CDSs with the best reciprocal BLAST *hit* in all the genomes used are taken into account.

3.8. *In silico* analysis of gene *clusters* of secondary metabolites and other putative bioactive products

To analyze gene *clusters* of secondary metabolites and other products of BAC3151, the *software* antiSMASH (MEDEMA et al., 2011) and BAGEL3 (VAN HEEL et al., 2013) were used. Insecticidal genes were searched for with BtToxin_scanner (YE et al., 2012). All programs were used with default parameters.

3.9. Nucleotide sequence access number

The *Whole Genome Shotgun* project has been deposited in DDBJ/EMBL /GenBank with the access number LDKD00000000. The version described in this book is LDKD01000000.

4. RESULTS AND DISCUSSION

4.1. Genomic features of the *B. thuringiensis* BAC3151 *draft* genome

The *draft* genome sequence of *B. thuringiensis* BAC3151 has 5,743,871 bp, distributed in 42 *scaffolds* (with an average size of 136,758 bp and N50 of 552.900 bp), and a total G+C content of 34.9% (A=1,867,003 bases [32.5%]; T=1,871,097 [32.57%]; C=1,008,063 [17.55%]; G=996,602 [17.35%]; N=1,106 [0.01%]). Table 2 shows the result of processing the *reads* until the *scaffolds* were assembled.

Table 2. Summary of the processing of data obtained from sequencing and assembly.

Initial number of *reads*	24.391.936
No. of filtered *reads* (≥Q20)	21.512.378
No. of *contigs*	112
No. of *scaffolds*	42

A total of 5562 protein coding sequences (CDSs) were predicted, with the CDSs comprising 4,604,274 bp, resulting in a protein coding percentage of approximately 80.2%. The G+C content of the CDSs was slightly higher than the global content and corresponded to 35.87% (A=1,596,789 bases [34.68%]; T=1,355,335 [29.43%]; C=689,264 [14.97%]; G=962,377 [20.9%]; N=509 [0.01%]). The gene density was 0.968 CDS/kb, with an average CDS size of 827 bp. The *draft* genome also contained 53 genes coding for tRNA and 4 for rRNA. The main characteristics of this and the other *B. thuringiensis* genomes used in this work are shown in Table 3.

B. thuringiensis strains have been shown to have similar G+C content, coding percentage and average CDS length. However, significant differences are observed in the size of the genome and also in the composition of strain-specific genes. Mobile genetic elements such as plasmids, transposons, phages, integrons and genomic islands play an important role in these differences (FROST et al., 2005; FANG et al., 2011; HE et al., 2011). In addition, 86 proteins from the predicted proteome of *B. thuringiensis* BAC315 did not generate significant alignment with the banks used (*e-value* <1e-6) and may be encoded by genes that have not yet been characterized. This number is higher than that of other *B. thuringiensis* strains with available genomes.

Table 3. General properties of *Bacillus thuringiensis* genomes used in this study.

Organization	*Status*	Size (Mb)	Access number	Reference
B. thuringiensis BAC3151	*Draft*	5,74	LDK00000000.1	This study
B. thuringiensis MC28	Complete	6,69	CP003687.1, CP003688.1 CP003693.1, CP003690.1 CP003689.1, CP003692.1 CP003691.1, CP003694.1	Guan et al. (2012)
B. thuringiensis Bt407	Complete	6,13	CP003889.1, CP003896.1 CP003893.1, CP003897.1 CP003890.1, CP003894.1 CP003891.1, CP003898.1 CP003895.1, CP003892.1	Sheppard et al. (2013)
B. thuringiensis BMB171	Complete	5,64	CP001903.1, CP001904.1	He et al. (2010)
B. thuringiensis HD- 789	Complete	6,33	CP003763.1, CP003769.1 CP003764.1, CP003766.1 CP003765.1, CP003768.1 CP003767.1	Doggett et al. (2013)
B. thuringiensis YBT-1518	Complete	6,67	CP005935.1, CP005937.1 CP005939.1, CP002486.1 CP005938.1, CP005940.1 CP005936.1	-
B. thuringiensis sv. *thuringiensis* str. IS5056	Complete	6,77	CP004123.1, CP004126.1 CP004127.1, CP004131.1 CP004128.1, CP004136.1 CP004137.1, CP004132.1 CP004135.1, CP004130.1 CP004134.1, CP004129.1 CP004124.1, CP004125.1 CP004133.1	Murawska et al. (2013)

Table 3. Continued.

Organization	*Status*	Size (Mb)	Access number	Reference
B. thuringiensis sv. *kurstaki* str. HD73	Complete	5,91	CP004069.1, CP004071.1 CP004072.1, CP004073.1 CP004070.1, CP004075.1 CP004076.1, CP004074.1	Liu et al. (2013)
B. thuringiensis sv. *konkukian* str. 97-27	Complete	5,31	AE017355.1, CP000047.1	Han et al. (2006)
B. thuringiensis str. Al	Complete	5,31	CP000485.1, CP000486.1	Challacombe et al. (2007)

B. thuringiensis sv. *finitimus* YBT-020	Complete	5,68	CP002508.1, CP002509.1, CP002510.1	Zhu et al. (2011)
B. thuringiensis sv. *chinensis* CT-43	Complete	6,15	CP001907.1, CP001913.1 CP001908.1, CP001917.1 CP001909.1, CP001915.1 CP001912.1, CP001910.1 CP001914.1, CP001916.1 CP001911.1	He et al. (2011)

4.2. Phylogenomic analysis

Conventional phylogenetic analysis markers, such as 16S rDNA, are essentially identical for *B. thuringiensis* strains (as well as others in the *Bacillus cereus* group) (BAVYKIN, et al., 2004) and the relationship between them using *housekeeping* genes is often inconsistent (HELGASON et al., 2004). For this reason, phylogenomic analysis was carried out using 1276 *core* genes from the strains (Figure 3).

BAC3151 was grouped together with *B. thuringiensis* MC28 isolated from soil and which is highly toxic to lepidopterans and dipterans (TAN et al., 2009; TAN et al., 2010). This result is consistent with the fact that soil is a potential source of endophytic bacteria (DOTY, 2008; RYAN et al., 2008). Other strains isolated from different environments were also grouped together. For example, strains obtained from soil, such as YBT-1518 and IS5056, grouped together with strains isolated from insects (ATCC 10792, T01001 [ZWICK et al., 2012] and Bt407).

Previous studies based on *multilocus sequence typing* (MLST) (PRIEST et al., 2004), BOX-PCR *fingerprinting* (KIM et al., 2002), genomic mapping (CARLSON et al., 1996), pulsed field gel electrophoresis (PFGE) (CARLSON et al., 1994), *multilocus* enzyme electrophoresis (MLEE) (HELGASON et. al., 2000), variable number of tandem repeats (VNTR), amplified fragment length polymorphism (AFLP) (TICKNOR et al., 2000), 2000), variable number of tandem repeats (VNTR) mapping, amplified fragment length polymorphism (AFLP) (TICKNOR et al., 2001) and mass spectrometry-based proteomic techniques integrated with statistical methods (DWORZANSKI et al., 2010), reveal the extensive similarity between *B. thuringiensis* strains as well as between other strains of the *B. cereus* group. The grouping of functionally distinct strains (as an insect pathogen and endophyte with antimicrobial activity) is in line with a previous study of the phylogenetic relationship between *B. thuringiensis* strains based on the complete genome (ZWICK et al. 2012).

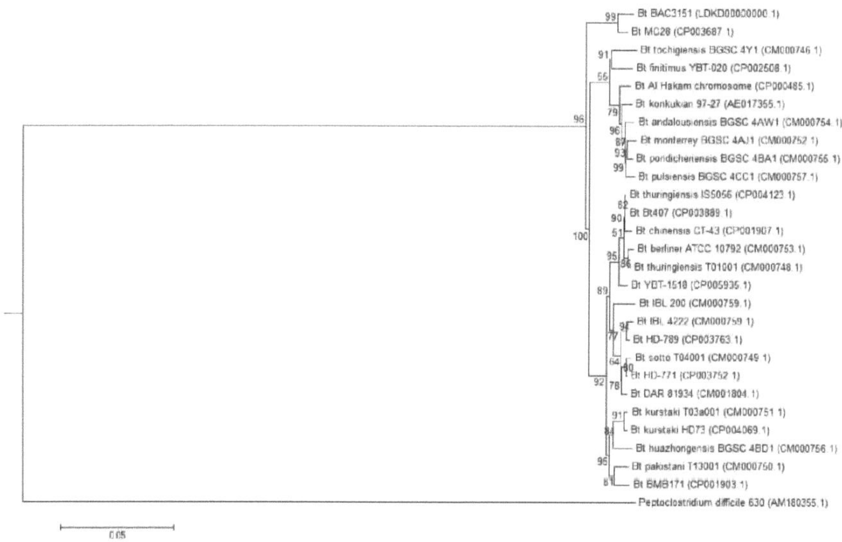

Figure 3. Phylogenomic analysis of *B. thuringiensis* based on 1276 *core* genes of the strains used. The tree was built with PHYLIP (FELSENSTEIN, 1989), using the NeighborJoining method with 1000 replicates. *Peptoclostridium difficile* 630 was used as an outgroup. *Bootstrap* values are shown on the nodes. The scale bar shows five nucleotide substitutions per 100 nucleotides. The accession number of the sequences used is given in brackets.

4.3. Arrangement and orientation of the *scaffolds*

B. thuringiensis MC28, which was the closest strain to BAC3151 (item 4.2), was used as a reference to determine the order and orientation of the *scaffolds* of our isolate. Table 4 shows the final position of the *scaffolds* (numbered in pseudo-coordinates) as well as their *forward* or *reverse* orientation. Despite the accuracy, some caution is needed when interpreting the results of the *scaffold* arrangements, since many true genomic rearrangements can occur in regions of repeated sequences, making it difficult to align the sequences. Rissman et al. (2009) reported an accuracy of between 90.4% and 99.4% for the ordering and orientation of sequences from different genome *drafts* using the *Mauve Contig Mover*.

Determining the order and orientation of the *scaffolds* facilitates the closure of *draft* genomes and comparative analyses. Using a reference genome that is more similar to the *draft* for this purpose produces a more accurate ordering and greater coverage of the sequences, as well as establishing their most likely orientation.

Table 4. Order and final orientation of all the *scaffolds* in the *B. thuringiensis* BAC3151 genomic *draft*.

Name	Guidance	Starting position	Final position
scaffold9	*reverse*	1	287073
scaffold4	*forward*	287074	351153
scaffold26[a]	*forward*	351154	1095058
scaffold24	*reverse*	1095059	1647958
scaffold16	*forward*	1647959	2063090
scaffold6[a]	*reverse*	2063091	2124415
scaffold32	*forward*	2124416	2225266
scaffold31	*forward*	2225267	2240649
scaffold3	*forward*	2240650	2891218
scaffold19	*forward*	2891219	3022029
scaffold20[a]	*forward*	3022030	3066338
scaffold23[a]	*reverse*	3066339	3080235
scaffold5	*reverse*	3080236	4082243
scaffold13[a]	*forward*	4082244	4506425
scaffold25	*forward*	4506426	4650005
C1123[a]	*reverse*	4650006	4664851
C1173[a]	*forward*	4664852	4713332

Table 4. Continued.

scaffold27	*reverse*	4713333	4771614
scaffold2	*forward*	4771615	4837291
scaffold18	*forward*	4837292	4858510
scaffold1[a]	*forward*	4858511	4887195
C1055	*reverse*	4887196	4889722
C1051	*reverse*	4889723	4892122
C1083[a]	*reverse*	4892123	4897873
C1105[a]	*reverse*	4897874	4905973
scaffold28	*reverse*	4905974	5043094
scaffold7	*reverse*	5043095	5125671
scaffold21	*forward*	5125672	5334211
scaffold 12	*reverse*	5334212	5523281
scaffold8	*forward*	5523282	5526453

scaffold10	*forward*	5526454	5554308
scaffold11	*forward*	5554309	5596095
scaffold 17	*forward*	5596096	5604454
scaffold22	*forward*	5604455	5611028
scaffold29	*forward*	5611029	5663233
scaffold30	*forward*	5663234	5703273
C1031	*forward*	5703274	5704386
C1035	*forward*	5704387	5705583
C1037	*forward*	5705584	5706865
C1079	*forward*	5706866	5712279
C1127	*forward*	5712280	5727434
C1131	*forward*	5727435	5743871

Scaffolds with possible alternative locations.

4.4. Gene sharing between strains

The pangenome, *core* genome *and* average number of *singletons of B. thuringiensis* were calculated based on the *draft* genome of BAC3151 and the 11 complete genomes described in table 3. The pangenome of the strains analyzed was estimated at 14755 CDSs, with 3413 *core* CDSs, which corresponds to approximately 49.3% to 71.1% of the CDSs in the genomes. The average number of *singletons* was 326 CDSs.

As more complete genome sequences of a species become available, the size of the pangenome of that species usually increases due to an increase in the number of accessory genes. Based on mathematical models, it is predicted that new genes will be discovered within the pangenome of many bacterial species even though hundreds or possibly thousands of complete genomic sequences have been characterized (MEDINI, 2005). Figure 4A shows the increase in the number of CDSs in the *B. thuringiensis* pangenome with the increase in the number of genomes analyzed. It can be seen that *B. thuringiensis* has an "open" pangenome. This observation is in line with the hypothesis that an "open" pangenome is the norm for the genus *Bacillus*, although *Bacillus anthracis* is an exception (FANG et al., 2011).

The reasons why only the 12 genomes were used instead of all the *B. thuringiensis* genomes available at NCBI for the calculations are twofold. Firstly, the pangenome of the species is open. Secondly, for the analyses of the *core* genome (Figure 4B) and *singletons* (Figure 4C), the curve remained practically the same with the addition of new genomes, so the analyses were only slightly biased. The analysis of the *core* genome showed that it conserved the CDSs of the general physiological and survival functions of the species. In a previous study, with only seven strains of *B. thuringiensis*, the proportion of genes shared between the strains varied within a smaller range (59.0% to 68.2%) than that observed by us (49.3% to 71.1%) (FANG et al., 2011). Figure 5 shows the sharing of CDSs of BAC3151 with the other *B. thuringiensis*. Our isolate has more CDSs in common with MC28 (4951 genes) and less with Al Hakam (3965 CDSs). This observation is consistent with the greater proximity of BAC3151 to MC28 observed in the phylogenetic analysis. The lower number of genes shared with Al Hakam may not only be the result of a less close relationship, but also because Al Hakam has a smaller gene repertoire, thus reducing the possibility of sharing.

In addition, there was a large variation in the number of *singletons* between *B. thuringiensis* strains, as reported by Fang et al. (2011). The terminology *singletons* describes the CDSs found in only one of the strains in our study. The same CDSs may be found in other strains or species. *B. thuringiensis* BAC3151 had 221 *singletons*, around 4% of the total number of CDSs, while in the other genomes this figure varied from around 0.4% to 11% (for the *chinensis* serovar CT-43 and Bt407, respectively). Most of the *singletons* were associated with hypothetical proteins, as observed in other strains and species (SMITS et al., 2010; JOSEPH et al., 2012). The large number of hypothetical proteins found in various lineages may be the result of genes originating from horizontal transfer, including from phylogenetically distant organisms, for which molecular characterization has not yet been carried out. In fact, *B. thuringiensis* can easily acquire genetic material from other bacteria, as previously reported (GONZALEZ et al., 1982; HAACK et al.,1996; VILAS-BÔAS et al.,1998). The acquisition of external genes can contribute to the better adaptation of subpopulations of bacteria to specific niches (LAING et al., 2010). Other *singletons* found were mainly involved in gene regulation, synthesis of cellular components, response to stimuli and processing of genetic information (replication and repair).

(A)

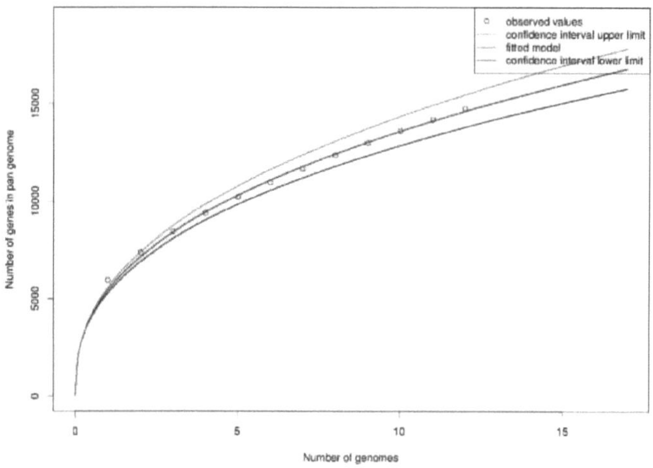
observed values
confidence interval upper limit
fitted model
confidence interval lower limit
Number of genes in pan genome
Number of genomes
0
5000
10000
15000
0
5
10
15

(B)

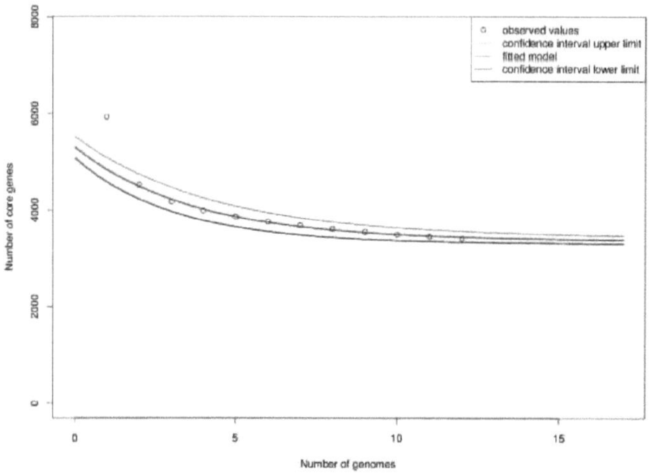
observed values
confidence interval upper limit
fitted model
confidence interval lower limit
Number of core genes
Number of genomes
0
2000
4000
6000
8000
0
5
10
15

(C)

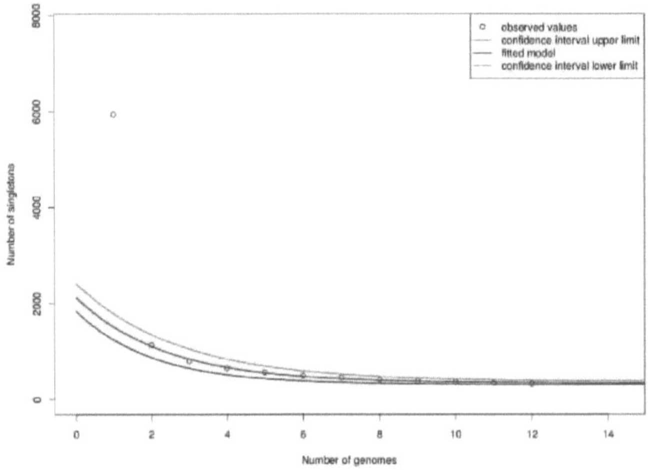

Figura 4. Development of the pangenome, *core* genome and *singletons* of *B. thuringiensis* with increasing number of genomes. Non-linear least squares fitting models predict the development of the pangenome (A), *core* genome (B) and singletons (C) of the genomes based on the functions $5422.261x^{0,399}$, $1935.165e^{(-x/3,664)} + 3367.738$ and $1793.351e^{(-x/2,355)} + 339.683$, respectively.

(A)

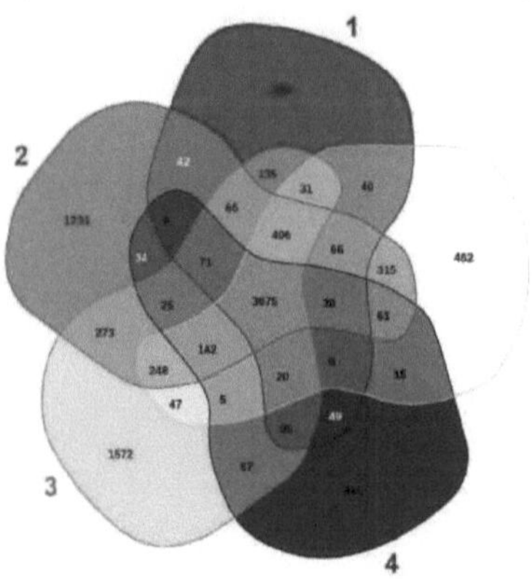

(B)

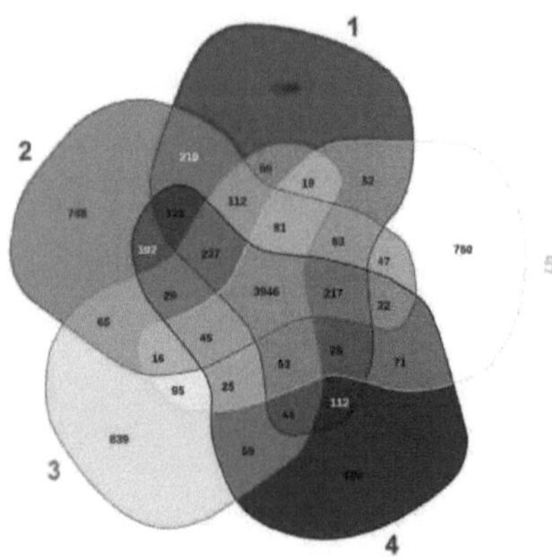

(C)

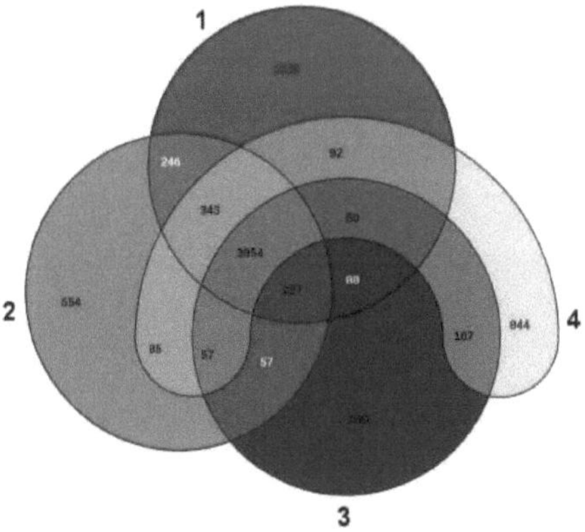

Figura 5. Venn diagrams constructed with EDGAR (BLOM et al., 2009) showing the sharing of BAC3151 CDSs with other genomes. (A) 1. BMB171, 2. MC28, 3. sv. *thuringiensis* IS5056, 4. Al Hakam, 5. BAC3151; (B) 1. YBT-1518, 2. sv. *chinensis* CT-43, 3. YBT-020, 4. sv. *kurstaki* HD73, 5. BAC3151; (C) 1. Bt407, 2. HD-789, 3. sv. *konkukian* 97-27, 4. BAC3151. Overlapping regions represent common CDSs between the genomes analyzed.

4.5. Putative genes of *B. thuringiensis* BAC3151 associated with antagonism

Microorganisms compete for limited resources within the community and the mechanisms of competition include antagonistic effects. *B. thuringiensis* BAC3151 has demonstrated antimicrobial activity *in vitro* against Gram-negative and Gram-positive bacteria, including *Curtobacterium flaccumfaciens* pv. *flaccumfaciens*, which is responsible for bacterial wilt of the bean plant (LOPES et al., 2015). Many gene *clusters* of secondary metabolites as well as genes of other putative products, which may be related to antimicrobial activity, have been identified in the *draft* genome of BAC3151. The main genes are shown in Table 5 and are potentially involved in the synthesis of bacteriocins, siderophores, polyketides, chitinases, N-acyl homoserine lactonase, insecticidal proteins and antibiotic compounds. These genes also point to possibilities for new tests that could broaden the potential use of BAC3151 for biological control of plant diseases.

Table 5. Main classes of key putative genes potentially involved in the antagonistic activity of *B. thuringiensis* BAC3151.

Class	Size (pb)	Product	Coverage	*E-value*	Similarity

Bacteriocins					
scaffold3_orf00655	68	Bacteriocin biosynthesis protein	0,985294	2,00E-30	1
scaffold21_orf00106	1435	Bacteriocin peptidase S8	0,987456	0	0,989422
scaffold26_orf00190	649	Bacteriocin biosynthesis protein (SagD)	0,998459	0	0,978428
scaffold29_orf00015	942	Lantibiotic biosynthesis protein (LanM)	0,998938	0	0,953291
Siderophores					
scaffold19_orf00032	253	Trilactone hydrolase (IroE)	0,996047	1,00E-136	0,948617
scaffold24_orf00456	612	Rhizobactin biosynthesis protein (RhbC)	0,998366	0	1
scaffold24_orf00453	327	Petrobactin biosynthesis protein (AsbE)	0,996942	0	1
scaffold24_orf00457	602	Petrobactin biosynthesis protein (AsbA)	0,998339	0	0,975083
Chitinases					
scaffold1_orf00008	235	Chitinase C	0,995745	1,00E-131	1

Table 5. Continued.

scaffold 19_orf00019	360	Chitinase	0,997222	0	0,977778
scaffold28_orf00015	688	Chitinase C	0,986919	0	0,954412
Polyketides					
scaffold9_orf00043	294	Polyketide synthase	0,996599	1,00E-171	1
scaffold 16_orf00424	47	Polyketide synthase	0,978723	1,00E-17	1
Insecticidal proteins					
C1127_orf00006	1584	Mosquitocidal protein	0,985	0	0,95
scaffold24_orf00006	292	δ-endotoxin	0,996575	1,00E-153	0,94863
Lactonase					
scaffold3_orf00391	250	N-acyl homoserine lactonase	0,996	1,00E-143	1
Antibiotics					
scaffold24_orf00119	206	Antibiotic biosynthesis monooxygenase	0,995146	1,00E-119	1
scaffold26_orf00002	311	Antibiotic biosynthesis monooxygenase	0,649518	1,00E-106	0,995074
scaffold32_orf00043	492	Antibiotic biosynthesis	0,457317	1,00E-108	0,884956

		monooxygenase			
scaffold32_orf00063	108	Antibiotic biosynthesis monooxygenase	0,990741	3,00E-55	1

Bacteriocins have been well described for *Bacillus*, especially for *B. subtilis*, although strains *of B. thuringiensis* also have the potential to produce them (READ et al., 2009). Although several bacteriocins from the *B. cereus* group have been reported, relatively few have been lantibiotic. In BAC3151, a lantibiotic *cluster* highly similar to the turicin biosynthetic *cluster* was found (Figure 6A). Turicin production has been described in different *B. thuringiensis* strains, such as turicin B439 from *B. thuringiensis* B439 (AHERN et al., 2003), turicin 17 from *B. thuringiensis* NEB17 (LEE et al., 2009b), turicin H from *B. thuringiensis* SF361 (LEE et al., 2009a) and turicin S from *B. thuringiensis* sv. *entomocidus* HD198 (CHEHIMI et al., 2007). This bacteriocin has shown antibacterial activity mainly against Gram-positive bacteria (FAVRET and YOUSTEN, 1989; REA et al., 2014). *Bacillus* bacteriocins are increasingly important due to their inhibition spectrum, which is sometimes broader than that of many lactic acid bacteria and can include Gram-negative bacteria, yeasts and other fungi, as well as Gram-positive bacteria (ABRIOUEL et al., 2011).

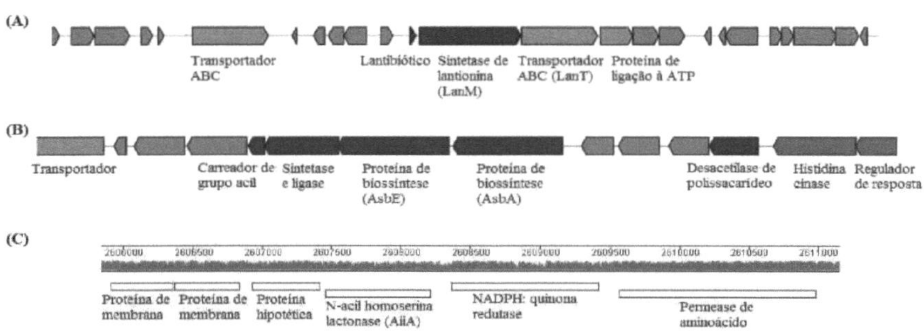

Figure 6. Some genomic regions of *B. thuringiensis* BAC3151 containing putative genes associated with microbial antagonism. (A) Lantibiotic gene *cluster* highly similar to that of turicin. (B) Petrobactin biosynthetic *cluster*. (C) Region containing the N-acyl homoserine lactonase gene, presumably involved in *quorum sensing* inhibition by BAC3151. In (A) and (B), biosynthesis, transport and regulatory genes are represented in red, blue and green, respectively. Other genes are represented in gray. In (C), the purple profile indicates the level of conservation of the region after alignment with other *B. thuringiensis* genomes.

Putative BAC3151 genes involved in siderophore biosynthesis include those present in the petrobactin biosynthetic *cluster* (Figure 6B). Petrobactin production has been reported in *Bacillus* species such as *B. thuringiensis* (WILSON et al., 2006), *B. cereus* (PARK et al., 2005) and *Bacillus anthracis* (CENDROWSKI et al., 2004). *B. thuringiensis* siderophores, such as petrobactin, may be especially important for biocontrol in agriculture, since *B. thuringiensis* is naturally associated with plants and competition for iron can lead to inhibition of phytopathogens, as well as increasing the supply of the element to the plant. Phytopathogens are generally not capable of producing comparable iron acquisition systems (RADDADI et al., 2007).

Another class of genes potentially involved in biocontrol is chitinases. Three putative *chitinase* genes (*chi* genes) have been identified in the genome. In *B. thuringiensis*, up to seven copies have been reported (MARCON et al., 2014). *Bacillus* spp. capable of degrading chitin in soil (KUZU et al., 2012), phylloplane (SMITT and COUCHE, 1991) and in association with insects (SNEH et al., 1993) have been described. Since chitin is an important component of the cell wall of fungi, chitinases can be used to control phytopathogenic fungi, protecting plants against diseases (TANG et al., 2012). Interestingly, the chitinolytic activity of *B. thuringiensis* may have a synergistic effect on insecticide toxicity. This can be applied to the development of better insecticidal products by combining both factors in one strain or in formulations (WIWAT et al., 2000; LIU et al., 2002).

In addition to these, genes for polyketide synthases were found. Due to the versatile synthesis mechanism of polyketides, they can exhibit diverse structures and a variety of biological functions, including antimicrobial activity. There is little information in the literature about *B. thuringiensis* polyketides. However, the growing number of genomes available for the species offers the opportunity to mine genes potentially involved in their synthesis and whose applicability could be explored. Polyketides, in fact, are difficult to obtain, which makes heterologous expression an alternative for use. Heterologous production systems can also facilitate the study of the catalytic properties of polyketide synthases. Examples of heterologous hosts for the production of these proteins include *Escherichia coli*, *B. subtilis* and *Streptomyces coelicolor*

(RUDE and KHOSLA, 2004).

Two insecticidal genes were also found in BAC3151. One (C1127_orf00006) with 95% similarity to a mosquitocidal protein gene from *B. thuringiensis* IBL 4222, and the other (scaffold24_orf00006) with 94.7% similarity to an insecticidal δ-endotoxin from *B. thuringiensis* MC28. The number of insecticidal genes varies between strains (FANG et al., 2011), which can produce one or more parasporal protein crystals. The individual insecticidal proteins that make up the crystals generally have a spectrum of toxic activity restricted to a few species of a particular order of insects, mainly lepidopterans, dipterans, coleopterans and hymenopterans (RHO et al., 2007). Activity against nematodes and other invertebrates has also been described (WEI et al., 2003; KOTZE et al., 2005; RHO et al., 2007). The combination of toxins from a given strain of *B. thuringiensis* defines the spectrum of activity of that strain. However, there are many crystal-producing strains with no known toxic activity (RHO et al., 2007). Further analysis could confirm the production of parasporal crystals by BAC3151 and its toxicity, as advances in the development of new biopesticides based on *B. thuringiensis* and the manipulation of its genes depend on the variety of strains available and their corresponding insecticidal genes.

With regard to the gene encoding N-acyl homoserine lactonase (gene *aiiA*) (Figure 6C), it is presumably responsible for the observed inhibition *of quorum sensing* in Gram-negative bacteria by BAC3151 (LOPES et al., 2015). Many Gram-negative phytopathogenic bacteria use *quorum sensing* as a mechanism for regulating different biological activities, including the production of virulence factors (VON BODMAN et al., 2003). *aiiA* has been found in different strains of *B. thuringiensis* (LEE et al., 2002) and inhibiting the chemical communication of phytopathogenic bacteria can be an efficient strategy for controlling plant diseases. Studies indicate that the use of *quorum sensing* inhibitors for this purpose is promising, expanding the current forms of bacterial disease prevention (DONG et al., 2004).

Other genes, especially those of monooxygenases, which possibly participate in the synthesis of antibiotic compounds were found in the genome of *B. thuringiensis*

BAC3151. These genes are often part of *clusters* involved in a complex secondary metabolism characteristic of strains producing various antimicrobials. Although the portion of the BAC3151 genome attributed to the synthesis of antimicrobials is apparently not as large as that of some *Bacillus* spp. such as B. *subtilis* and *B. amyloliquefaciens* (KUNST et al., 1997; CHEN et al., 2009), future research should analyze the functionality of the genes found, the conservation of the synthesis pathways and the properties of the eventual products as well as their applicability in agriculture or even in other areas. In this way, the isolate could represent a source of secondary metabolites and enzymes that are important for controlling pathogenic microorganisms.

5. CONCLUSIONS

In this work, it was found that *B. thuringiensis* BAC3151 endophytic from common bean was more similar to *B. thuringiensis* MC28 isolated from soil. The genomic analyses also revealed several BAC3151 genes without significant similarity to others in the public databases as well as genes not found in the other strains analyzed and which will be better investigated in the future. In addition, putative genes involved in multiple forms of antagonism are present. In summary, the results presented from the genomic analyses could contribute to the development of new strategies for the biological control of diseases of bean and other crops.

6. REFERENCES

ABDERRAHMANI, A.; TAPI, A.; NATECHE, F.; CHOLLET, M.; LECLERE, V.; WATHELET, B.; et al. Bioinformatics and molecular approaches to detect NRPS genes involved in the biosynthesis of kurstakin from *Bacillus thuringiensis*. **Appl Microbiol Biotechnol**, v. 92, p. 571-581, 2011.

ABRIOUEL, H.; FRANZ, C. M.; BEN OMAR, N.; GALVEZ, A. Diversity and applications of *Bacillus* bacteriocins. **FEMS Microbiol Rev**, v. 35, p. 201-232, 2011.

AHERN, M.; VERSCHUEREN, S.; VAN SINDEREN, D. Isolation and characterization of a novel bacteriocin produced by *Bacillus thuringiensis* strain B439. **FEMS Microbiol Lett**, v. 220, p. 127-131, 2003.

AKIBA, Y. Microbial ecology of *Bacillus thuringiensis*. **Appl Ent Zool**, v. 21, p. 76-80, 1986.

ALKAN, C.; SAJJADIAN, S.; EICHLER, E. E. Limitations of next-generation genome sequence assembly. **Nat Methods**, v. 8, p. 61-65, 2011.

ALTSCHUL, S. F.; GISH, W.; MILLER, W.; MYERS, E. W.; LIPMAN, D. J. Basic local alignment search tool. **J Mol Biol**, v. 215, p. 403-410, 1990.

ANDREOLLI, M.; LAMPIS, S.; POLI, M.; GULLNER, G.; BIRO, B.; VALLINI, G. Endophytic *Burkholderia fungorum* DBT1 can improve phytoremediation efficiency of polycyclic aromatic hydrocarbons. **Chemosphere**, v. 92, p. 688-694, 2013.

ANSARI, M. Z.; YADAV, G.; GOKHALE, R. S.; MOHANTY, D. NRPS- PKS: a knowledge-based resource for analysis of NRPS/PKS megasynthases. **Nucleic Acids Res**, v. 32, p. W405-413, 2004.

ARANTES, O. M. N.; VILAS-BÔAS, L. A.; VILAS-BÔAS, G. F. L. T. *Bacillus thuringiensis*: strategies in biological control. In: SERAFINI, L. A.; BARROS, N. M.; AZEVEDO, J. L. (org). **Biotecnologia na agricultura e agroindùstria**, Guaiba: Agropecuària, v. 2, p. 269-293, 2002.

ARGUELLES-ARIAS, A.; ONGENA, M.; HALIMI, B.; LARA, Y.; BRANS, A.;

JORIS, B.; et al. *Bacillus amyloliquefaciens* GA1 as a source of potent antibiotics and other secondary metabolites for biocontrol of plant pathogens. **Microb Cell Fact**, v. 8, p. 63, 2009.

ARONSON, A. I.; BECKMAN, W.; DUNN, P. *Bacillus thuringiensis* and related insect pathogens. **Microbiol Rev**, v. 50, p. 1-24, 1986.

BAKER, M. *De novo* genome assembly: what every biologist should know. **Nat Meth**, v. 9, p. 333-337, 2012.

BAKKER, P. A. H. M.; PIETERSE, C. M. J.; VAN LOON, L. C. Induced systemic resistance by fluorescent *Pseudomonas* spp. **Phytopathology**, v. 97, p. 239243, 2007.

BANSODE, V. B.; BAJEKAL, S. S. Characterization of chitinases from microorganisms isolated from Lonar Lake. **Ind J Biotech**, v. 5, p. 357-363, 2006.

BARBOZA-CORONA, J. E.; VAZQUEZ-ACOSTA, H.; BIDESHI, D. K.; SALCEDO-HERNANDEZ, R. Bacteriocin-like inhibitor substances produced by Mexican strains *of Bacillus thuringiensis*. **Arch Microbiol**, v. 187, p. 117-126, 2007.

BARKA, E. A.; GOGNIES, S.; NOWAK, J.; AUDRAN, J. C.; BELARBI, A. Inhibitory effect of endophyte bacteria on *Botrytis cinerea* and its influence to promote the grapevine growth. **Biol Control**, v. 24, p. 135-142, 2002.

BAVYKIN, S. G.; LYSOV, Y. P.; ZAKHARIEV, V.; KELLY, J. J.; JACKMAN, J.; STAHL, D. A.; et al. Use of 16S rRNA, 23S rRNA, and *gyrB* gene sequence analysis to determine phylogenetic relationships of *Bacillus cereus* group microorganisms. **J Clin Microbiol**, v. 42, p. 3711-3730, 2004.

BENDTSEN, J. D.; NIELSEN, H.; VON HEIJNE, G.; BRUNAK, S. Improved prediction of signal peptides: SignalP 3.0. **J Mol Biol**, v. 340, p. 783-795, 2004.

BERLINER, E. Eber die schlaffsucht der Mehlmottenraupe. (*Ephesis kuehniella* Zell.) undihren Erreger *Bacillus thuringiensis*, n. sp. Z. ang. **Entomology**, v. 2, p. 29-56, 1915.

BERLINER, E. Über die schlafsucht der Mehlmottenraupe. **Z. Gesante**

Getreidewesen, v. 3, p. 63-70, 1911.

BHATTACHRYA, D.; NAGPURE, A.; GUPTA, R. K. Bacterial chitinase: properties and potential. **Critic Rev Biotechnol**, v. 27, p. 21-28, 2007.

BLOM, J.; ALBAUM, S. P.; DOPPMEIER, D.; PUHLER, A.; VORHOLTER, F. J.; ZAKRZEWSKI, M; et al. EDGAR: a software framework for the comparative analysis of prokaryotic genomes. **BMC Bioinformatics**, v. 10, p. 154, 2009.

BODE, H.B. Entomopathogenic bacteria as a source of secondary metabolites. **Curr Opin Chem Biol**, v. 13, p. 2224-230, 2009.

BRADER, G.; COMPANT, S.; MITTER, B.; TROGNITZ, F.; SESSITSCH, A. Metabolic potential of endophytic bacteria. **Curr Opin Biotechnol**, v. 27, p. 3037, 2014.

BRAR, S. K.; VERMA, M.; TYAGI, R. D.; VALÉRO, J. R. Recent advances in downstream processing and formulations of *Bacillus thuringiensis* based biopesticides. **Process Biochem**, v. 41, p. 323-342, 2006.

BRAVO, A.; SARABIA, S.; LOPEZ, L.; ONTIVEROS, H.; ABARCA, C.; ORTIZ, A.; et al. Characterization of *cry* genes in a Mexican *Bacillus thuringiensis* strain collection. **Appl Environ Microbiol**, v. 64, p. 4965-4972, 1998.

CARDENAS, M. I.; GALAN-WONG, L.; FERRÉ-MANZANERO, J.; PEREYRA-ALFÉREZ, B. Selection of *cry* toxins against *Thichoplusia ni* . **Cienc Uanl**, v. 4, p. 51-62, 2001.

CARLSON, C. R.; CAUGANT, D. A.; KOLSTO, A. B. Genotypic diversity among *Bacillus cereus* and *Bacillus thuringiensis* strains. **Appl Environ Microbiol**, v. 60, p. 1719-1725, 1994.

CARLSON, C. R.; JOHANSEN, T.; KOLSTO, A. B. The chromosome map of *Bacillus thuringiensis* subsp. *canadensis* HD224 is highly similar to that of the *Bacillus cereus* type strain ATCC 14579. **FEMS Microbiol Lett**, v. 141, p. 163-167, 1996.

CASTRESANA J. Selection of conserved blocks from multiple alignments for their use in phylogenetic analysis. **Mol Biol Evol**, v. 17, p. 540-552, 2000.

CENDROWSKI, S.; MACARTHUR, W.; HANNA, P. *Bacillus anthracis* requires siderophore biosynthesis for growth in macrophages and mouse virulence. **Mol Microbiol**, v. 51, p. 407-417, 2004.

CHAISSON, M.; PEVZNER, P. Short read fragment assembly of bacterial genomes. **Genome Res**, v. 18, p. 324, 2008.

CHALLACOMBE, J. F.; ALTHERR, M. R.; XIE, G.; BHOTIKA, S. S.; BROWN, N.; BRUCE, D.; et al. The complete genome sequence of *Bacillus thuringiensis* Al Hakam. **J Bacteriol**, v. 189, p. 3680-3681, 2007.

CHEHIMI, S.; DELALANDE, F.; SABLE, S.; HAJLAOUI, M. R.; VAN DORSSELAER, A.; LIMAM, F.; et al. Purification and partial amino acid sequence of thuricin S, a new *anti-Listeria* bacteriocin from *Bacillus thuringiensis*. **Can J Microbiol**, v. 53, p. 284-290, 2007.

CHEN, X. H.; KOUMOUTSI, A.; SCHOLZ, R.; EISENREICH, A.; SCHNEIDER, K.; HEINEMEYER, I.; et al. Comparative analysis of the complete genome sequence of the plant growth-promoting bacterium *Bacillus amyloliquefaciens* FZB42. **Nat Biotechnol**, v. 25, p. 1007-1014, 2007.

CHEN, X. H.; KOUMOUTSI, A.; SCHOLZ, R.; SCHNEIDER, K.; VATER, J.; SUSSMUTH, R.; et al. Genome analysis of *Bacillus amyloliquefaciens* FZB42 reveals its potential for biocontrol of plant pathogens. **J Biotechnol**, v. 140, p. 27-37, 2009.

CHERIF, A.; CHEHIMI, S.; LIMEM, F.; HANSEN, B. M.; HENDRIKSEN, N. B.; DAFFONCHIO, D.; et al. Detection and characterization of the novel bacteriocin entomocin 9, and safety evaluation of its producer, *Bacillus thuringiensis* ssp. *entomocidus* HD9. **J Appl Microbiol**, v. 95, p. 990-1000, 2003.

CHERIF, A.; REZGUI, W.; RADDADI, N.; DAFFONCHIO, D.; BOUDABOUS, A. Characterization and partial purification of entomocin 110, a newly identified bacteriocin from *Bacillus thuringiensis* subsp. *entomocidus* HD110. **Microbiol Res**, v. 163, p. 684-692, 2008.

CHEVREUX, B.; PFISTERER, T.; DRESCHER, B.; DRIESEL, A. J.; MULLER, W. E.; WETTER, T.; et al. Using the miraEST assembler for reliable and automated mRNA transcript assembly and SNP detection in sequenced ESTs. **Genome Res**, v. 14, p. 1147-1159, 2004.

COOMBS, J. T.; MICHELSEN, P. P.; FRANCO, C. M. M. Evaluation of endophytic actinobacteria as antagonists *of Gaeumannomyces graminis* var. *tritici* in wheat. **Biol Control**, v. 29, p. 359-366, 2004.

COSTA, L. E. O.; QUEIROZ, M. V.; BORGES, A. C.; MORAES, C. A.; ARAÙJO, E. F. Isolation and characterization of endophytic bacteria isolated from the leaves of the common bean (*Phaseolus vulgaris*). **Braz J Microbiol**, v. 43, p. 1562-1575, 2012.

CRICKMORE, N.; BAUM, J.; BRAVO, A.; LERECLUS, D.; NARVA, K.; SAMPSON, K.; et al. ***Bacillus thuringiensis* toxin nomenclature**. Available at: http://www.btnomenclature.info/. Accessed on: December 20, 2014.

DAHIYA, N.; TEWARI, R.; HOONDAL, G. S. Biotechnological aspects of chitinolytic enzymes: a review. **Appl Microbiol Biotechnol**, v. 71, p. 773-782, 2006.

D'ALESSANDRO, M.; ERB, M.; TON, J.; BRANDENBURG, A.; KARLEN, D.; ZOPFI, J.; et al. Volatiles produced by soil-borne endophytic bacteria increase plant pathogen resistance and affect tritrophic interactions. **Plant Cell Environ**, v. 37, p. 813-826, 2014.

DARLING, A. C.; MAU, B.; BLATTNER, F. R.; PERNA, N. T. Mauve: multiple alignment of conserved genomic sequence with rearrangements. **Genome Res**, v. 14, p. 1394-1403, 2004.

DE LA FUENTE-SALCIDO, N.; GUADALUPE ALANIS-GUZMAN, M.; BIDESHI, D. K.; SALCEDO-HERNANDEZ, R.; BAUTISTA-JUSTO, M.; BARBOZA-CORONA, J. E. Enhanced synthesis and antimicrobial activities of bacteriocins produced by Mexican strains *ofBacillus thuringiensis*. **Arch Microbiol**, v. 190, p. 633-640, 2008.

DE LA VEGA, L. M.; BARBOZA-CORONA, J. E.; AGUILAR-USCANGA, M. G.;

RAMIREZ-LEPE, M. Purification and characterization of an exochitinase from *Bacillus thuringiensis* subsp. *aizawai* and its action against phytopathogenic fungi. **Can J Microbiol**, v. 52, p. 651-657, 2006.

DE MAAGD, R. A.; BOSCH, D.; STIEKEMA, W. *Bacillus thuringiensis* toxin-mediated insect resistance in plants. **Trends Plant Sci**, v. 4, p. 9-13, 1999.

DELCHER, A. L.; HARMON, D.; KASIF, S.; WHITE, O.; SALZBERG, S. L. Improved microbial gene identification with GLIMMER. **Nucleic Acids Res**, v. 27, p. 4636-4641, 1999.

DIDELOT, X.; FALUSH, D. Inference of bacterial microevolution using multilocus sequence data. **Genetics**, v. 175, p. 1251-1266, 2007.

DOGGETT, N. A.; STUBBEN, C. J.; CHERTKOV, O.; BRUCE, D. C.; DETTER, J. C.; JOHNSON, S. L.; HAN, C. S. Complete genome sequence of *Bacillus thuringiensis* serovar *israelensis* strain HD-789. **Genome Announc**, v. 1, p. 1-2, 2013.

DOHM, J. C.; LOTTAZ, C.; BORODINA, T.; HIMMELBAUER, H. SHARCGS, a fast and highly accurate short-read assembly algorithm for de novo genomic sequencing. **Genome Res**, v. 17, p. 1697-1706, 2007.

DONG, Y. H.; ZHANG, X. F.; XU, J. L.; ZHANG, L. H. Insecticidal *Bacillus thuringiensis* silences *Erwinia carotovora* virulence by a new form of microbial antagonism, signal interference. **Appl Environ Microbiol**, v. 70, p. 954-960, 2004.

DOTY, S. L. Enhancing phytoremediation through the use of transgenics and endophytes. **New Phytol,** v. 179, p. 318-333, 2008.

DWORZANSKI, J. P.; DICKINSON, D. N.; DESHPANDE, S. V.; SNYDER, A. P.; ECKENRODE, B. A. Discrimination and phylogenomic classification of *Bacillus anthracis-cereus-thuringiensis* strains based on LC-MS/MS analysis of whole cell protein digests. **Anal Chem**, v. 82, p. 145-155, 2010.

EDGAR, R. C. MUSCLE: multiple sequence alignment with high accuracy and high throughput. **Nucleic Acids Res**, v. 32, p. 1792-1797, 2004.

EICHLER, E. E.; SANKOFF, D. Structural dynamics of eukaryotic chromosome evolution. **Science**, v. 301, p. 793-797, 2003.

EKBLOM, R.; WOLF, J. B. A field guide to whole-genome sequencing, assembly and annotation. **Evol Appl**, v. 7, p. 1026-1042, 2014.

FANG, Y.; LI, Z.; LIU, J.; SHU, C.; WANG, X.; ZHANG, X.; et al. A pangenomic study of *Bacillus thuringiensis*. **J Genet Genomics**, v. 38, p. 567-576, 2011.

FAVRET, M. E.; YOUSTEN, A. A. Thuricin: The bacteriocin produced by *Bacillus thuringiensis*. **J Invertebr Pathol**, v. 53, p. 206-216, 1989.

FEDERICI, B. A. *Bacillus thuringiensis* in Biological Control. In: BELLOWS, T.; FISHER, T. W. (eds). **Handbook of Biological Control**. California: University of California, p. 575-593, 1999.

FEITELSON, J. S. Novel pesticidal delta-endotoxins from *Bacillus thuringiensis*. In: **Proceedings of XXVII Annual Meeting of the Society for Invertebrate Pathology**, Montpellier, France, p. 184, 1994.

FELNAGLE, E. A.; JACKSON, E. E.; CHAN, Y. A.; PODEVELS, A. M.; BERTI, A. D.; MCMAHON, M. D.; et al. Nonribosomal peptide synthetases involved in the production of medically relevant natural products. **Mol Pharm**, v. 5, p. 191-211, 2008.

FELSENSTEIN, J. PHYLIP: Phylogeny Inference Package (version3.2). **Cladistics**, v. 5, p. 164-166, 1989.

FERREIRA, L. H. P. L.; SUZUKI, M. T.; ITANO, E. N.; ONO, M. A.; ARANTES, O. M. N. Ecological aspects of *Bacillus thuringiensis* in an Oxisol. **Sci Agric**, v. 60, p.19-22, 2003.

FICKERS, P. Antibiotic compounds from *Bacillus*: Why are they so amazing? **Am J Biochem Biotechnol**, v. 8, p. 38-43, 2012.

FLEISCHMANN, R. D.; ADAMS, M. D.; WHITE, O.; CLAYTON, R. A.; KIRKNESS, E. F.; KERLAVAGE, A. R.; et al. Whole-genome random sequencing and assembly of *Haemophilus influenzae* Rd. **Science**, v. 269, p. 496-512, 1995.

FRASER, C. M.; GOCAYNE, J. D.; WHITE, O.; ADAMS, M. D.; CLAYTON, R. A.; FLEISCHMANN, R. D.; et al. The minimal complement gene of *Mycoplasma genitalium*. **Science**, v. 270, p. 397-403, 1995.

FROMMEL, M. I.; NOWAK, J.; LAZAROVITS, G. Growth enhancement and developmental modifications of *in vitro* grown potato (*Solanum tuberosum* spp. *tuberosum*) as affected by a nonfluorescent *Pseudomonas* sp. **Plant Physiol**, v. 96, p.928-936, 1991.

FROST, L. S.; LEPLAE, R.; SUMMERS, A. O.; TOUSSAINT, A. Mobile genetic elements: the agents of open source evolution. **Nat Rev Microbiol**, v. 3, p. 722-732, 2005.

FUQUA, C.; PARSEK, M. R.; GREENBERG, E. P. Regulation of gene expression by cell-to-cell communication: acyl-homoserine lactone quorum sensing. **Annu Rev Genet**, v. 35, p. 439-468, 2001.

GALLEGOS, M. T.; SCHLEIF, R.; BAIROCH, A.; HOFMANN, K.; RAMOS, J. L. Arac/XylS family of transcriptional regulators. **Microbiol Mol Biol Rev**, v. 61, p. 393-410, 1997.

GARDNER, J. M.; FELDMAN, A. W.; ZABLOTOWICZ, R. M. Identity and behavior of xylem-residing bacteria in rough lemon roots of florida citrus trees. **Appl Environ Microbiol**, v. 43, p. 1335-1342, 1982.

GELERNTER, W.; SCHWAB, G. E. Transgenic bacteria, viruses, algae and other microorganisms as *Bacillus thuringiensis* toxin delivery systems. In: ENTWISTLE, P. F.; CORY, J. S.; BAILEY, M. J.; HIGGS, S. (ed). ***Bacillus thuringiensis*, an environmental biopesticide: theory and practice**, Chichester: John Wiley & Sons, p. 89-104, 1993.

GISH, W.; STATES, D. J. Identification of protein coding regions by database similarity search. **Nat Genet**, v. 3, p. 266-272, 1993.

GLARE, T. R; O'CALLAGHAN, M. *Bacillus thuringiensis*: biology, ecology and safety. In: ENTWISTLE, P. F.; CORY, J. S.; BAILEY, M. J.; HIGGS, S. (ed). ***Bacillus***

***thuringiensis*, an environmental biopesticide: theory and practice**, Chichester: John Wiley, p. 350, 2000.

GOMAA, E. Z. Chitinase Production by *Bacillus thuringiensis* and *Bacillus licheniformis*: their potential in antifungal biocontrol. **J Microbiol**, v. 50, p. 103-111, 2012.

GONZALEZ, J. M., JR.; BROWN, B. J.; CARLTON, B. C. Transfer of *Bacillus thuringiensis* plasmids coding for delta-endotoxin among strains *of B. thuringiensis* and *B. cereus*. **Proc Natl Acad Sci USA**, v. 79, p. 6951-6955, 1982.

GORDON, D.; ABAJIAN, C.; GREEN, P. Consed: a graphical tool for sequence finishing. **Genome Res**, v. 8, p. 195-202, 1998.

GRIFFITHS-JONES, S.; MOXON, S.; MARSHALL, M.; KHANNA, A.; EDDY, S. R.; BATEMAN, A. Rfam: annotating non-coding RNAs in complete genomes. **Nucleic Acids Res**, v. 33, p. D121-124, 2005.

GUAN, P.; AI, P.; DAI, X.; ZHANG, J.; XU, L.; ZHU, J.; et al. Complete genome sequence of *Bacillus thuringiensis* serovar *sichuansis* strain MC28. **J Bacteriol**, v. 194, p. 6975, 2012.

GUNDOGDU, O.; BENTLEY, S. D.; HOLDEN, M. T.; PARKHILL, J.; DORRELL, N.; WREN, B. W. Re-annotation and re-analysis of the *Campylobacter jejuni* NCTC11168 genome sequence. **BMC Genomics**, v. 8, p. 162, 2007.

GUSTAFSON, K. R.; CARDELLINA, J. H., 2ND; FULLER, R. W.; WEISLOW, O. S.; KISER, R. F.; SNADER, K. M.; et al. AIDS-antiviral sulfolipids from cyanobacteria (blue-green algae). **J Natl Cancer Inst**, v. 81, p. 1254-1258, 1989.

HAACK, B. J.; ANDREWS, R. E.; LOYNACHAN, T. E. Tn916-mediated genetic exchange in soil. **Soil Biol Biochem**, v. 28, p. 765-771, 1996.

HALLMANN, J.; QUADT-HALLMANN, A.; MAHAFEE, W. F.; KLOEPPER, J. Bacterial endophytes in agricultural crops. **Can J Microbiol**, v. 43, p. 895-914, 1997.

HALLMANN, J.; QUADT-HALLMANN, A.; RODRIGUEZ-KABANA, R.;

KLOEPPER, J. W., Interactions between *Meloidogyne incognita* and endophytic bacteria in cotton and cucumber. **Soil Biol Biochem**, v. 30, p. 925-937, 1998.

HAN, C. S.; XIE, G.; CHALLACOMBE, J. F.; ALTHERR, M. R.; BHOTIKA, S. S.; BROWN, N.; et al. Pathogenomic sequence analysis of *Bacillus cereus* and *Bacillus thuringiensis* isolates closely related to *Bacillus anthracis*. **J Bacteriol**, v. 188, p. 3382-3390, 2006.

HANNAY, C. L.; FITZ-JAMES, P. C. The protein-crystals *of Bacillus thuringiensis berliner*. **Can J Microbiol**, v. 1, p. 694-710, 1995.

HANSON, A. D.; PRIBAT, A.; WALLER, J. C.; DE CRECY-LAGARD, V. 'Unknown' proteins and 'orphan' enzymes: the missing half of the engineering parts list-and how to find it. **Biochem J**, v. 425, p. 1-11, 2010.

HARDOIM, P. R.; VAN OVERBEEK, L. S.; ELSAS, J. D. Properties of bacterial endophytes and their proposed role in plant growth. **Trends Microbiol**, v. 16, 463-471, 2008.

HAYES, M.; CARNEY, B.; SLATER, J.; BRUCK, W. Mining marine shellfish wastes for bioactive molecules: chitin and chitosan - Part A: extraction methods. **Biotechnol J**, v. 3, p. 871-877, 2008.

HE, J.; SHAO, X.; ZHENG, H.; LI, M.; WANG, J.; ZHANG, Q.; et al. Complete genome sequence of *Bacillus thuringiensis* mutant strain BMB171. **J Bacteriol**, v. 192, p. 4074-4075, 2010.

HE, J.; WANG, J.; YIN, W.; SHAO, X.; ZHENG, H.; LI, M.; et al. Complete genome sequence of *Bacillus thuringiensis* subsp. *chinensis* strain CT-43. **J Bacteriol**, v. 193, p. 3407-3408, 2011.

HELGASON, E.; OKSTAD, O. A.; CAUGANT, D. A.; JOHANSEN, H. A.; FOUET, A.; MOCK, M.; et al. *Bacillus anthracis*, *Bacillus cereus*, and *Bacillus thuringiensis* - one species on the basis of genetic evidence. **Appl Environ Microbiol**, v. 66, p. 2627-2630, 2000.

HELGASON, E.; TOURASSE, N. J.; MEISAL, R.; CAUGANT, D. A.; KOLSTO, A.

B. Multilocus sequence typing scheme for bacteria of the *Bacillus cereus* group. **Appl Environ Microbiol**, v. 70, p. 191-201, 2004.

HENDRIKSEN, N. B.; HANSEN, B. M. Long-term survival and germination of *Bacillus thuringiensis* var. *kurstaki* in a field trial. **Can J Microbiol**, v. 48, p. 256261, 2002.

HERTWECK, C.; LUZHETSKYY, A.; REBETS, Y.; BECHTHOLD, A. Type II polyketide synthases: gaining a deeper insight into enzymatic teamwork. **Nat Prod Rep**, v. 24, p. 162-190, 2007.

HUANG, J. Ultrastructure of bacterial penetration in plants. **Annu Rev Phytopathol**, v. 24, p. 141-157, 1986.

HUNT, M.; NEWBOLD, C.; BERRIMAN, M.; OTTO, T. D. A comprehensive evaluation of assembly scaffolding tools. **Genome Biol**, v. 15, p. R42, 2014.

HUNTER, S.; APWEILER, R.; ATTWOOD, T. K.; BAIROCH, A.; BATEMAN, A.; BINNS, D.; et al. InterPro: the integrative protein signature database. **Nucleic Acids Res**, v. 37, p. D211-215, 2009.

HUREK, T.; REINHOLD-HUREK, B.; VAN MONTAGU, M.; KELLENBERGER, E. Root colonization and systemic spreading of *Azoarcus* sp. strain BH72 in grasses. **J Bacteriol**, v. 176, 1913-1923, 1994.

HUSZ, B. On the use of *Bacillus thuringiensis* in the fight against the corner borer. **Instit Corn Borer Invest Sci Repub**, v.2, p. 99-110, 1929.

IBRAHIM, M. A.; GRIKO, N.; JUNKER, M.; BULLA, L. A. *Bacillus thuringiensis*: a genomics and proteomics perspective. **Bioeng Bugs**, v. 1, p. 31-50, 2010.

ISHIWATA, S. On a kind of severe flacherie (sotto disease). **Danihon Sanshi Kaiho**, v. 114, p. 1-5, 1901.

JENKE-KODAMA, H.; DITTMANN, E. Evolution of metabolic diversity: insights from microbial polyketide synthases. **Phytochemistry**, v. 70, p. 1858-1866, 2009.

JEONG, H.; PARK, S. H.; CHOI, S. K. Genome sequence of the acrystalliferous

Bacillus thuringiensis serovar *israelensis* strain 4Q7, widely used as a recombination host. **Genome Announc**, v. 2, p. e00231-14, 2014.

JHA, P. N.; KUMAR, A. Endophytic colonization of *Typha australis* by a plant growth-promoting bacterium *Klebsiella oxytoca* strain GR-3. **J Appl Microbiol**, v. 103, p. 1311-1320, 2007.

JONES, P.; BINNS, D.; CHANG, H. Y.; FRASER, M.; LI, W.; MCANULLA, C.; et al. InterProScan 5: genome-scale protein function classification. **Bioinformatics**, v. 30, p. 1236-1240, 2014.

JOSEPH, S.; DESAI, P.; JI, Y.; CUMMINGS, C. A.; SHIH, R.; et al. Comparative analysis of genome sequences covering the seven *Cronobacter* species. **Plos One**, v. 7, p. e49455, 2012.

KIM, W.; HONG, Y. P.; YOO, J. H.; LEE, W. B.; CHOI, C. S.; CHUNG, S. I. Genetic relationships of *Bacillus anthracis* and closely related species based on variable-number tandem repeat analysis and BOX-PCR genomic *fingerprinting*. **FEMS Microbiol Lett**, v. 207, p. 21-27, 2002.

KLOEPPER, J. W.; RYU, C. M., ZHANG, S. Induced systemic resistance and promotion of plant growth by *Bacillus* spp. **Phytopathology**, v. 94, p. 1259-1266, 2004.

KOBAYASHI, T.; SUZUKI, M.; INOUE, H.; ITAI, R. N.; TAKAHASHI, M.; NAKANISHI, H.; et al. Expression of iron-acquisition-related genes in iron-deficient rice is co-ordinately induced by partially conserved iron-deficiency-responsive elements. **J Exp Bot**, v. 56, p. 1305-1316, 2005.

KOONIN, E. V. Computational genomics. **Curr Biol**, v. 11, p. R155-158, 2001.

KOTZE, A. C.; O'GRADY, J.; GOUGH, J. M.; PEARSON, R.; BAGNALL, N. H.; KEMP, D. H.; et al. Toxicity of Bacillus thuringiensis to parasitic and free- living life-stages of nematode parasites of livestock. **Int J Parasitol**, v. 35, p. 10131022, 2005.

KUNST, F.; OGASAWARA, N.; MOSZER, I.; ALBERTINI, A. M.; ALLONI, G.; AZEVEDO, V.; et al. The complete genome sequence of the grampositive bacterium

Bacillus subtilis. **Nature**, v. 390, p. 249-256, 1997.

KUZU, S. B.; GUVENMEZ, H. K.; DENIZCI, A. A. Production of a thermostable and alkaline chitinase by *Bacillus thuringiensis* subsp. *kurstaki* strain HBK-51. **Biotechnol Res Int**, v. 2012, p. 135498, 2012.

KYRPIDES, N. C. Fifteen years of microbial genomics: meeting the challenges and fulfilling the dream. **Nat Biotechnol**, v. 27, p. 627-632, 2009.

LAGESEN, K.; HALLIN, P.; RODLAND, E. A.; STAERFELDT, H. H.; ROGNES, T.; USSERY, D. W. RNAmmer: consistent and rapid annotation of ribosomal RNA genes. **Nucleic Acids Res**, v. 35, p. 3100-3108, 2007.

LAING, C.; BUCHANAN, C.; TABOADA, E. N.; ZHANG, Y.; KROPINSKI, A.; VILLEGAS, A.; et al. Pan-genome sequence analysis using Panseq: an online tool for the rapid analysis of core and accessory genomic regions. **BMC Bioinformatics**, v. 11, p. 461, 2010.

LAMB, T. G.; TONKYN, D. W.; KLUEPFEL, D. A. Movement of *Pseudomonas aureofaciens* from the rhizosphere to aerial plant tissue. **Can J Microbiol**, v. 42, p. 1112-1120, 1996.

LASLETT, D.; CANBACK, B. ARAGORN, a program to detect tRNA genes and tmRNA genes in nucleotide sequences. **Nucleic Acids Res**, v. 32, p. 11-16, 2004.

LEE, H.; CHUREY, J. J.; WOROBO, R. W. Biosynthesis and transcriptional analysis of thurincin H, a tandem repeated bacteriocin genetic locus, produced by *Bacillus thuringiensis* SF361. **FEMS Microbiol Lett**, v. 299, p. 205-213, 2009a.

LEE, K. D.; GRAY, E. J.; MABOOD, F.; JUNG, W. J.; CHARLES, T.; CLARK, S. R.; et al. The class IId bacteriocin thuricin-17 increases plant growth. **Planta**, v. 229, p. 747-755, 2009b.

LEE, S. J.; PARK, S. Y.; LEE, J. J.; YUM, D. Y.; KOO, B. T.; LEE, J. K. Genes encoding the N-acyl homoserine lactone-degrading enzyme are widespread in many subspecies of *Bacillus thuringiensis*. **Appl Environ Microbiol**, v. 68, p. 39193924, 2002.

LIBRADO, P.; ROZAS, J. DnaSP v5: a software for comprehensive analysis of DNA polymorphism data. **Bioinformatics**, v. 25, p. 1451-1452, 2009.

LIU, G.; SONG, L.; SHU, C.; WANG, P.; DENG, C.; PENG, Q.; et al. Complete genome sequence of *Bacillus thuringiensis* subsp. *kurstaki* strain HD73. **Genome Announc**, v. 1, p. e0008013, 2013.

LIU, M.; CAI, Q. X.; LIU, H. Z.; ZHANG, B. H.; YAN, J. P.; YUAN, Z. M. Chitinolytic activities in *Bacillus thuringiensis* and their synergistic effects on larvicidal activity. **J Appl Microbiol**, v. 93, p. 374-379, 2002.

LIU, X.; ZHOU, R.; FU, G.; ZHANG, W.; MIN, Y.; TIAN, Y.; et al. Draft genome sequence of *Bacillus thuringiensis* NBIN-866 with high nematocidal activity. **Genome Announc**, v. 2, p.1-2, 2014.

LOPES, R. B. M; COSTA, LEO; VANETTI, M. C. D; ARAUJO E. F.; QUEIROZ M. V. Endophytic bacteria isolated from common bean (*Phaseolus vulgaris*) exhibiting high variability showed antimicrobial activity and quorum sensing inhibition. **Curr Microbiol**, v. 71, p. 509-516, 2015.

LÓPEZ-LÓPEZ, A.; ROGEL, M. A.; ORMENO-ORRILLO, E.; MARTiNEZ-ROMERO, J.; MARTiNEZ-ROMERO, E. *Phaseolus vulgaris* seed-borne endophytic community with novel bacterial species such as *Rhizobium endophyticum* sp. nov. **Syst Appl Microbiol**, v. 33, p. 322-327, 2010.

LOWE, T. M.; EDDY, S. R. tRNAscan-SE: a program for improved detection of transfer RNA genes in genomic sequence. **Nucleic Acids Res**, v. 25, p. 955-964, 1997.

LUKASHIN, A. V.; BORODOVSKY, M. GeneMark.hmm: new solutions for gene finding. **Nucleic Acids Res**, v. 26, p. 1107-1115, 1998.

LUO, C.; HU, G.; ZHU, H. Genome reannotation of *Escherichia coli* CFT073 with new insights into virulence. **BMC Genomics**, v. 10, p. 552, 2009.

MACLEAN, D.; JONES, J. D.; STUDHOLME, D. J. Application of 'nextgeneration' sequencing technologies to microbial genetics. **Nat Rev Microbiol**, v. 7, p. 287-296, 2009.

MADUPU, R.; BRINKAC, L. M.; HARROW, J.; WILMING, L. G.; BOHME, U.; LAMESCH, P.; et al. Meeting report: a workshop on best practices in genome annotation. **Database (Oxford)**, v. 2010, p. baq001, 2010.

MARCHLER-BAUER, A.; LU, S.; ANDERSON, J. B.; CHITSAZ, F.; DERBYSHIRE, M. K.; DEWEESE-SCOTT, C.; et al. CDD: a Conserved Domain Database for the functional annotation of proteins. **Nucleic Acids Res**, v. 39, p. D225-229, 2011.

MARCO, G.; MANUEL, P. Ecological mysteries: Is *Bacillus thuringiensis* a real insect pathogen? **Bt Res.**, v.3, p. 1-2, 2012.

MARCON, J.; TAKETANI, R. G.; DINI-ANDREOTE, F.; MAZZERO, G. I.; SOARES, F. L. J.; MELO, I. S.; et al. Draft genome sequence of *Bacillus thuringiensis* strain BrMgv02-JM63, a chitinolytic bacterium isolated from oil-contaminated mangrove soil in Brazil. **Genome Announc**, v. 2, p. 2014.

MARDIS, E. R. The impact of next-generation sequencing technology on genetics. **Trends Genet**, v. 24, p. 133-141, 2008.

MARTIN, R. G.; ROSNER, J. L. The AraC transcriptional activators. **Curr Opin Microbiol**, v. 4, p. 132-137, 2001.

MARTiNEZ, L.; CABALLERO-MELLADO, J.; OROZCO, J.; MARTiNEZ-ROMERO, E. Diazotrophic bacteria associated with banana (*Musa* spp.). **Plant Soil**, v. 257, p. 35-47, 2003.

MARTiNEZ-ROMERO, E. Coevolution in *Rhizobium* -legume symbiosis? **DNA Cell Biol.**, v. 28, p. 361-370, 2009.

MARTiNEZ-ROMERO, E. Diversity of *Rhizobium-Phaseolus vulgaris* symbiosis: overview and perspectives. **Plant Soil**, v. 252, p. 11-23, 2003.

MCINROY, J. A.; KLOEPPER, J. W. Population dynamics of endophytic bacteria in field-grown sweet corn and cotton. **Can J Microbiol**, v. 41, p. 895-901, 1995.

MEADOWS, M. P. *Bacillus thuringiensis* in the environment: ecology and risk assessment. In: ENTWISTLE, P. F.; CORY, J. S.; BAILEY, M. J.; HIGGS, S. (ed). ***Bacillus thuringiensis*** **an environment biopesticide: theory and practice**, Chichester: John Wiley & Sons, p. 193-220, 1993.

MEDEMA, M. H.; BLIN, K.; CIMERMANCIC, P.; DE JAGER, V.; ZAKRZEWSKI, P.; FISCHBACH, M. A.; et al. Antismash: Rapid identification, annotation and analysis of secondary metabolite biosynthesis gene clusters in bacterial and fungal genome sequences. **Nucleic Acids Res**, v. 39, p.339-346, 2011.

MÉDIGUE, C.; MOSZER, I. Annotation, comparison and databases for hundreds of bacterial genomes. **Res Microbiol**, v. 158, p. 724-736, 2007.

MEDINI, D.; DONATI, C.; TETTELIN, H.; MASIGNANI, V.; RAPPUOLI, R. The microbial pan-genome. **Curr Opin Genet Dev**, v. 15, p. 589-594, 2005.

MEURER, G.; GERLITZ, M.; WENDT-PIENKOWSKI, E.; VINING, L. C.; ROHR, J.; HUTCHINSON, C. R. Iterative type II polyketide synthases, cyclases and ketoreductases exhibit context-dependent behavior in the biosynthesis of linear and angular decapolyketides. **Chem Biol**, v. 4, p. 433-443, 1997.

MISAGHI, I. J.; DONNDELINGER, C. R. Endophytic bacteria in symptom- free cottton plants. **Phytopathology**, v. 80, p. 808-811, 1990.

MISHRA, P. K.; MISHRA, S.; SELVAKUMAR, G.; BISHT, J. K.; KUNDU, S.; GUPTA, H.S. Coinoculation of *Bacillus thuringiensis-KR1* with *Rhizobium leguminosarum* enhances plant growth and nodulation of pea (*Pisum sativum* L.) and lentil (*Lens culinaris* L.). **World J Microbiol Biotechnol**, v. 25, p. 753-761, 2009.

MISHRA, R.; SINGH, R.; JAISWAL, H.; KUMAR, V.; MAURYA, S. *Rhizobium-mediated* induction of phenolics and plant growth promotion in rice (*Oryza sativa* L.). **Curr Microbiol**, v. 52, p. 383-389, 2006.

MONNERAT, R. G.; SANTOS, R.; BARROS, P.; BATISTA, A.; BERRY, C. **Isolation and characterization of endophytic *Bacillus thuringiensis* strains from cotton**, Brasilia: Embrapa Recursos Genéticos e Biotecnologia, 4 p., 2003.

MURAWSKA, E.; FIEDORUK, K.; BIDESHI, D. K.; SWIECICKA, I. Complete genome sequence of *Bacillus thuringiensis* subsp. *thuringiensis* strain IS5056, an isolate highly toxic to *Trichoplusia ni*. **Genome Announc**, v. 1, p. e0010813, 2013.

NAVON, A. *Bacillus thuringiensis* application in agriculture. In: CHARLES, J. F.; DELÉCLUSE, A.; NILESEN-LE ROUX, C. (ed.). **Entomopathogenic bacteria: from laboratory to field application**, Netherlands: Kluwer academic Publishers, p. 355-367, 2000.

OVERBEEK, R.; BEGLEY, T.; BUTLER, R. M.; CHOUDHURI, J. V.; CHUANG, H. Y.; COHOON, M.; et al. The subsystems approach to genome annotation and its use in the project to annotate 1000 genomes. **Nucleic Acids Res**, v. 33, p. 5691-5702, 2005.

PAIK, H. D.; BAE, S. S.; PARK, S. H.; PAN, J. G. Identification and partial characterization of tochicin, a bacteriocin offduced by *Bacillus thuringiensis* subsp. *tochigiensis*. **J Ind Microbiol Biotechnol**, v. 19, p. 294-298, 1997.

PALMA, L.; MUNOZ, D.; BERRY, C.; MURILLO, J.; CABALLERO, P. Draft genome sequences of two *Bacillus thuringiensis* strains and characterization of a putative 41.9-kDa insecticidal toxin. **Toxins (Basel**), v. 6, p. 1490-1504, 2014.

PARK, R. Y.; CHOI, M. H.; SUN, H. Y.; SHIN, S. H. Production of catecholsiderophore and utilization of transferrin-bound iron in *Bacillus cereus*. **Biol Pharm Bull**, v. 28, p. 1132-1135, 2005.

PATEL, P. S.; HUANG, S.; FISHER, S.; PIRNIK, D.; AKLONIS, C.; DEAN, L.; et al. Bacillaene, a novel inhibitor of prokaryotic protein synthesis produced by Bacillus subtilis: production, taxonomy, isolation, physico-chemical characterization and biological activity. **J Antibiot (Tokyo)**, v. 48, p. 997-1003, 1995.

PAULSEN, I. T.; PRESS, C. M.; RAVEL, J.; KOBAYASHI, D. Y.; MYERS, G. S.; MAVRODI, D. V.; et al. Complete genome sequence of the plant commensal *Pseudomonas fluorescens* Pf-5. **Nat Biotechnol**, v. 23, p. 873-878, 2005.

POLANCZYK, R; ALVES, S. *Bacillus thuringiensis*: a brief review. **Agrociência**, v. 8, p. 1-10, 2003.

PRAÇA, L. B.; GOMES, A. C. M. M.; CABRAL, G.; MARTINS, E. S.; SUJII, E. R.; MONNERAT, R. G. Endophytic colonization by brazilian strains of *Bacillus thuringiensis* on cabbage seedlings grown *in vitro*. **Bt Res**, v. 3, p. 11-19, 2012.

PRICE, A. L.; JONES, N. C.; PEVZNER, P. A. De novo identification of repeat families in large genomes. **Bioinformatics**, v. 21, p. i351-358, 2005.

PRIEST, F. G.; BARKER, M.; BAILLIE, L. W.; HOLMES, E. C.; MAIDEN, M. C. Population structure and evolution of the *Bacillus cereus* group. **J Bacteriol**, v. 186, p. 7959-7970, 2004.

QUADT-HALLMANN, A.; KLOEPPER, J. W. Immunological detection and localization of the cotton endophyte *Enterobacter asburiae* JM22 in different plant species. **Can J Microbiol**, v. 42, p. 1144-1154, 1996.

QUINONES, B.; DULLA, G.; LINDOW, S. E. Quorum sensing regulates exopolysaccharide production, motility and virulence in *Pseudomonas syringae.* **Mol Plant Microbe Interact**, v 18, p. 682-693, 2005.

RADDADI, N; CHERIF, A.; OUZARI, H.; MARZORATI, M.; BRUSETTI, L.; et al. *Bacillus thuringiensis* beyond insect biocontrol: plant growth promotion and biosafety of polyvalent strains, **Ann Microbiol**, v. 57, p. 481-494, 2007.

RASKO, D. A.; ROSOVITZ, M. J.; MYERS, G. S.; MONGODIN, E. F.; FRICKE, W. F.; GAJER, P.; et al. The pangenome structure of *Escherichia coli*: comparative genomic analysis of *E. coli* commensal and pathogenic isolates. **J Bacteriol**, v. 190, p. 6881-6893, 2008.

REA, M. C.; ALEMAYEHU, D.; CASEY, P. G.; O'CONNOR, P. M.; LAWLOR P. G.; WALSH. M.; et al. Bioavailability of the anti-clostridial bacteriocin thuricin CD in gastrointestinal tract. **Microbiology**, v. 160, p. 439-445, 2014.

READ, T. D.; AKMAL, A.; BISHOP-LILLY, K. **Annotation of the *Bacillus thuringiensis* BGSC4Y1 genome**. Available at:

http://www.ncbi.nlm.nih.gov/nuccore/NZ CM000746.1, 2009.

REESE, M. G. Application of a time-delay neural network to promoter annotation in the *Drosophila melanogaster* genome. **Comput Chem**, v. 26, p. 51-56, 2001.

REEVES, G. A.; TALAVERA, D.; THORNTON, J. M. Genome and proteome annotation: organization, interpretation and integration. **J R Soc Interface**, v. 6, p. 129-147, 2009.

RISSMAN, A. I.; MAU, B.; BIEHL, B. S.; DARLING, A. E.; GLASNER, J. D.; PERNA, N. T. Reordering contigs of draft genomes using the Mauve aligner. **Bioinformatics**, v. 25, p. 2071-2073, 2009.

ROH, J. Y.; CHOI, J. Y.; LI, M. S.; JIN, B. R.; JE, Y. H. Bacillus thuringiensis as a specific, safe, and effective tool for insect pest control. **J Microbiol Biotechnol**, v. 17, p. 547-559, 2007.

ROJAS-AVELIZAPA, L. I.; CRUZ-CAMARILLO, R.; GUERRERO, M. I.; RODRIGUEZ-VAZQUEZ, R.; IBARRA, J. E. Selection and characterization of a proteo-chitinolytic strain of *Bacillus thuringiensis*, able to grow in waste media. **World J Microbiol Biotechnol**, v. 15, p. 261-268, 1999.

ROULI, L.; M, M. B.; ROBERT, C.; NDIAYE, M.; LA SCOLA, B.; RAOULT, D. Genomic analysis of three African strains *of Bacillus anthracis* demonstrates that they are part of the clonal expansion of an exclusively pathogenic bacterium. **New Microbes New Infect**, v. 2, p. 161-169, 2014.

ROSEN BLUETH, M.; MARTINEZ-ROMERO, E. Bacterial endophytes and their interactions with hosts. **Mol Plant Microbe Interact**, v. 19, p. 827-837, 2006.

RUDE, M. A; KHOSLA, C. Engineered biosynthesis of polyketides in heterologous hosts. **Chem Eng Sci**, v. 59, p. 4693-4701, 2004.

RUST, A. G.; MONGIN, E.; BIRNEY, E. Genome annotation techniques: new approaches and challenges. **Drug Discov Today**, v. 7, p. S70-76, 2002.

RUTHERFORD, K.; PARKHILL, J.; CROOK, J.; HORSNELL, T.; RICE, P.; RAJANDREAM, M. A.; et al. Artemis: sequence visualization and annotation. **Bioinformatics**, v. 16, p. 944-945, 2000.

RYAN, R. P.; GERMAINE, K.; FRANKS, A.; RYAN, D. J.; DOWLING, D. N. Bacterial endophytes: recent developments and applications. **FEMS Microbiol Lett**, v. 278, p. 1-9, 2008.

SADFI, N.; CHERIF, M.; FLISS, I.; BOUDABBOUS, A.; ANTOUN, H. Evaluation of *Bacillus* isolates from salty soils and *Bacillus thuringiensis* strains for the biocontrol of *Fusarium* dry rot of potato tubers. **J Plant Pathol**, v. 83, p. 101-118, 2001.

SALAMA, H. S.; MORRIS, O. N. The use of *Bacillus thuringiensis* in developing countries. In: ENTWISTLE, P. F.; CORY, J. S.; BAILEY, M. J.; HIGGS, S. (ed). ***Bacillus thuringiensis*, an environmental biopesticide: theory and practice**, Chichester: John Wiley & Sons, p. 237-253, 1993.

SALZBERG, S. L. Genome re-annotation: a wiki solution? **Genome Biol**, v. 8, p. 102, 2007.

SANGER, F.; AIR, G. M.; BARRELL, B. G.; BROWN, N. L.; COULSON, A. R.; FIDDES, C. A.; et al. Nucleotide sequence of bacteriophage phi X174 DNA. **Nature**, v. 265, p. 687-695, 1977a.

SANGER, F.; NICKLEN, S.; COULSON, A. R. DNA sequencing with chain-terminating inhibitors. **Proc Natl Acad Sci USA**, v. 74, p. 5463-5467, 1977b.

SCHNEPF, E.; CRICKMORE, N.; VAN RIE, J.; LERECLUS, D.; BAUM, J.; FEITELSON, J.; et al. *Bacillus thuringiensis* and its pesticidal crystal proteins. **Microbiol Mol Biol Rev**, v. 62, p. 775-806, 1998.

SCOTT, B. Epichloë endophytes: fungal symbionts of grasses. **Curr Opin Microbiol**, v. 4, p. 393-398, 2001.

SHEPPARD, A. E.; POEHLEIN, A.; ROSENSTIEL, P.; LIESEGANG, H.; SCHULENBURG, H. Complete genome sequence of *Bacillus thuringiensis* strain 407 *cry-*. **Genome Announc**, v. 1, p. 1-2, 2013.

SMITH, R. A; COUCHE, G. A. The phylloplane as a source of *Bacillus thuringiensis* variants. **Appl Environ Microbiol**, v. 57, p. 311-315, 1991.

SMITH, R. A; COUCHE, G. A. The phylloplane as a source of *Bacillus thuringiensis* variants. **Appl Environ Microbiol**, v. 57, p. 311-315, 1991.

SMITS, T. H.; JAENICKE, S.; REZZONICO, F.; KAMBER, T.; GOESMANN, A.; FREY, J. E.; et al. Complete genome sequence of the fire blight pathogen *Erwinia pyrifoliae* DSM 12163T and comparative genomic insights into plant pathogenicity. **BMC Genomics**, v. 11, p. 2, 2010.

SNEH, B.; SCHUSTER S.; GROSS, S. Improvement of the insecticidal activity of *Bacillus thuringiensis* var. *entomocidus* on larvae *of Spodoptera fittoralis* (Lepidoptera, Noctuidae) by addition of chitinolytic bacteria, a phagostimulant and a UV-protectant. **Z Angew Entomol**, v. 96, p. 77-83, 1983.

SPRENT, J.; FARIA, S. Mechanisms of infection of plants by nitrogen fixing organisms. **Plant Soil**, v. 110, p. 157-165, 1988.

STAHLY, P.; ANDREWS, R. E.; YOUSTEN, A. A. The genus *Bacillus* - Insect Pathogens. In: BALOWS, A. H. G.; TRUPPER, M.; DWORKIN, W; HERDER, K.H. **The Prokaryotes**, 2nd ed., New York: Springer-Verlag, p. 4770, 1991.

STRAGIER, P.; KUNKEL, B.; KROOS, L.; LOSICK, R. Chromosomal rearrangement generating a composite gene for a developmental transcription factor. **Science**, v. 243, p. 507-512, 1989.

STEIN, L. Genome annotation from sequence to biology. **Nat Rev**, v. 2, p. 493-505, 2001.

STUDHOLME, D. J.; DIXON, R. Domain architectures of sigma54-dependent transcriptional activators. **J Bacteriol**, v. 185, p. 1757-1767, 2003.

SUBRAHMANYAN, P.; REDDY, M. N.; RAO, A. S. Exudation of certain organic compounds from seeds of groundnut. **Seed Sci Technol**, v. 11, p. 267-27, 1983.

SUGINTA, W.; ROBERTSON, P. A.; AUSTIN, B.; FRY, S. C.; FOTHERGILL-GILMORE, L. A. Chitinases from Vibrio: activity screening and purification of *chiA* from *Vibrio carchariae*. **J Appl Microbiol**, v. 89, p. 76-84, 2000.

SULLIVAN, T. J.; RODSTROM, J.; VANDOP, J.; LIBRIZZI, J.; GRAHAM, C.; SCHARDL, C. L., et al. Symbiont-mediated changes in *Lolium arundinaceum* inducible defenses: evidence from changes in gene expression and leaf composition. **New Phytol**, v. 176, p. 673-679, 2007.

SUN, Y.; ZHAO, Q.; XIA, L.; DING, X.; HU, Q.; FEDERICI, B. A.; et al. Identification and characterization of three previously undescribed crystal proteins from *Bacillus thuringiensis* subsp. *jegathesan*. **Appl Environ Microbiol**, v. 79, p. 3364-3370, 2013.

SUZUKI, M. T.; HERNANDEZ-RODRiGUEZ, C. S.; ARAÙJO, W. L.; FERRÉ, J. Characterization of an endophytic *Bacillus thuringiensis* strain isolated from sugar cane. In: **Proceedings of 41st Annual Meeting of the Society for Invertebrate Pathology and 9th International Conference on *Bacillus thuringiensis***, Coventry, United Kingdom, August 3-7, 2008.

SUZUKI, M. T.; LERECLUS, D.; ARANTES, O. M. N. Fate of *Bacillus thuringiensis* strains in different insect larvae. **Can J Microbiol**, v. 50, p. 973-975, 2004.

TAMURA, K.; STECHER, G.; PETERSON, D.; FILIPSKI, A.; KUMAR, S. MEGA6: Molecular Evolutionary Genetics Analysis version 6.0. **Mol Biol Evol**, v. 30, p. 2725-2729, 2013.

TAN, F.; ZHENG, A.; ZHU, J.; WANG, L.; LI, S.; DENG, Q.; WANG, S.; LI, P.; TANG, X. Rapid cloning, identification, and application of one novel crystal protein gene *cry30Fa1* from *Bacillus thuringiensis*. **FEMS Microbiol Lett**, v. 302, p. 46-51, 2010.

TAN, F.; ZHU, J.; TANG, J.; TANG, X.; WANG, S.; ZHENG, A.; LI, P. Cloning and characterization of two novel crystal protein genes, *cry54Aa1* and *cry30Fa1*, from *Bacillus thuringiensis* strain BtMC28. **Curr Microbiol**, v. 58, p. 654-659, 2009.

TANG, Y.; ZOU, J.; ZHANG, L.; LI, Z.; MA, C.; MA, N. Anti-fungi activities of *Bacillus thuringiensis* H3 chitinase and immobilized chitinase particles and their effects to rice seedling defensive enzymes. **J Nanosci Nanotechnol**, v. 12, p.

80818086, 2012.

TANUJA, R.; BISHT, S. C.; MISHRA, P. K. Ascending migration of endophytic *Bacillus thuringiensis* and assessment of benefits to different legumes of N.W. Himalayas. **Eur J Soil Biol**, v. 56, p. 56-64, 2012.

TETTELIN, H.; MASIGNANI, V.; CIESLEWICZ, M. J.; DONATI, C.; MEDINI, D.; WARD, N. L.; et al. Genome analysis of multiple pathogenic isolates of *Streptococcus agalactiae*: implications for the microbial "pan-genome". **Proc Natl Acad Sci USA**, v. 102, p. 13950-13955, 2005.

THOMAS, D. J. I.; ALUN, J.; MORGAN, W.; WHIPPS, J. M.; SAUNDERS, J. R. Plasmid transfer between the *Bacillus thuringiensis* subspecies *kurstaki* and *tenebrionis* in laboratory culture and soil and in Lepidopteran and Coleopteran larvae. **Appl Environ Microbiol**, v. 66, p. 118-124, 2000.

TICKNOR, L. O.; KOLSTO, A. B.; HILL, K. K.; KEIM, P.; LAKER, M. T.; TONKS, M.; et al. Fluorescent amplified fragment length polymorphism analysis of Norwegian *Bacillus cereus* and *Bacillus thuringiensis* soil isolates. **Appl Environ Microbiol**, v. 67, p. 4863-4873, 2001.

TILLIER, E. R.; COLLINS, R. A. Genome rearrangement by replication- directed translocation. **Nat Genet**, v. 26, p. 195-197, 2000.

TURNBAUGH, P. J.; LEY, R. E.; HAMADY, M.; FRASER-LIGGETT, C. M.; KNIGHT, R.; GORDON, J. I. The human microbiome project. **Nature**, v. 449, p. 804-810, 2007.

VAN FRANKENHUZEN, K. The challenge of *Bacillus thuringiensis*. In: ENTWISTLE, P. F.; CORY, J. S.; BAILEY, M. J.; HIGGS, S. (ed). ***Bacillus thuringiensis*, an environment biotpesticide: theory and practice**, Chichester: John Wiley & Sons, p. 1-35. 1993.

VAN HEEL, A. J; DE JONG, A.; MONTALBAN-LOPEZ, M.; KOK, J.; KUIPERS, O. P. BAGEL3: Automated identification of genes encoding bacteriocins and (non-) bactericidal posttranslationally modified peptides. **Nucleic Acids Res**, v. 41, p. 448-

453, 2013.

VAN LANEN, S. G.; SHEN, B. Microbial genomics for the improvement of natural product discovery. **Curr Opin Microbiol**, v. 9, p. 252-260, 2006.

VENTER, J. C.; ADAMS, M. D.; MYERS, E. W.; LI, P. W.; MURAL, R. J.; SUTTON, G. G.; et al. The sequence of the human genome. **Science**, v. 291, p. 1304-1351, 2001.

VILAS-BÔAS, G. F. L. T.; VILAS-BÔAS, L. A.; LERECLUS, D.; ARANTES, O. M. N. *Bacillus thuringiensis* conjugation under environmental conditions. **FEMS Microbiol Ecol**, v. 25, p. 369-374, 1998.

VILAS-BÔAS, L. A.; VILAS-BÔAS, G. F. L. T.; SARIDAKIS, H. O.; LEMOS, M. V. F.; LERECLUS, D.; ARANTES, O. M. N. Survival and conjugation of *Bacillus thuringiensis* in a soil microcosm. **FEMS Microbiol Ecol**, v. 31, p. 225259, 2000.

VON BODMAN, S. B.; BAUER, W. D.; COPLIN, D. L. Quorum sensing in plant-pathogenic bacteria. **Annu Rev Phytopathol**, v. 41, p. 455-482, 2003.

WANG, A.; PATTEMORE, J.; ASH, G.; WILLIAMS, A.; HANE, J. Draft genome sequence of *Bacillus thuringiensis* strain DAR 81934, which exhibits molluscicidal activity. **Genome Announc**, v. 1, p. 1-2, 2013.

WANG, S. L.; HWANG, J. Microbial reclamation of shellfish wastes for the production of chitinases. **Enzyme Microb Technol**, v. 28, p. 376-382, 2001.

WANG, S. L.; LIN, T. Y.; YEN, Y. H.; LIAO, H. F.; CHEN, Y. J. Bioconversion of shellfish chitin wastes for the production of *Bacillus subtilis* W-118 chitinase. **Carbohydr Res**, v. 341, p. 2507-2515, 2006.

WEI, J. Z.; HALE, K.; CARTA, L.; PLATZER, E.; WONG, C.; FANG, S. C.; et al. Bacillus thuringiensis crystal proteins that target nematodes. **Proc Natl Acad Sci U S A**, v. 100, p. 2760-2765, 2003.

WEISER, J. Impact of *Bacillus thuringiensis* on applied entomology in eastern Europe and in Soviet Union. In: KRIEG, A.; HUGER, A. **M. Mitteilungen aus der**

biologischen bundesanstalt für land und forstwirtschaft Berlin-Dahlem heft, Berlin: Paul Parey, p. 37-50, 1986.

WILSON, K. E.; FLOR, J. E.; SCHWARTZ, R. E.; JOSHUA, H.; SMITH, J. L.; PELAK, B. A.; et al. Difficidin and oxydifficidin: novel broad spectrum antibacterial antibiotics produced by *Bacillus subtilis*. II. Isolation and physicochemical characterization. **J Antibiot (Tokyo)**, v. 40, p. 1682-1691, 1987.

WILSON, M. K.; ABERGEL, R. J.; RAYMOND, K. N.; ARCENEAUX, J. E.; BYERS, B. R. Siderophores of *Bacillus anthracis*, *Bacillus cereus*, and *Bacillus thuringiensis*. **Biochem Biophys Res Commun**, v. 348, p. 320-325, 2006.

WIWAT, C.; THAITHANUN, S.; PANTUWATANA, S.; BHUMIRATANA, A. Toxicity of chitinase-producing Bacillus *thuringiensis* ssp. *kurstaki* HD-1 (G) toward *Plutella xylostella*. **J Invertebr Pathol**, v. 76, p. 270-277, 2000.

YE, W.; ZHU, L.; LIU, Y.; CRICKMORE, N.; PENG, D.; RUAN, L.; et al. Mining new crystal protein genes from *Bacillus thuringiensis* on the basis of mixed plasmid-enriched genome sequencing and a computational pipeline. **Appl Environ Microbiol**, v. 78, p. 4795-4801, 2012.

ZAWADZKA, A. M.; KIM, Y.; MALTSEVA, N.; NICHIPORUK, R.; FAN, Y.; JOACHIMIAK, A.; et al. Characterization of a *Bacillus subtilis* transporter for petrobactin, an anthrax stealth siderophore. **Proc Natl Acad Sci USA**, v. 106, p. 21854-21859, 2009.

ZGHAL, R. Z.; JAOUA, S. Evidence of DNA rearrangements in the 128- kilobase pBtoxis plasmid of *Bacillus thuringiensis israelensis*. **Mol Biotechnol**, v. 33, p. 191-198, 2006.

ZHANG, G.; DENG, A.; XU, Q.; LIANG, Y.; CHEN, N.; WEN, T. Complete genome sequence of *Bacillus amyloliquefaciens* TA208, a strain for industrial production of guanosine and ribavirin. **J Bacteriol**, v. 193, p. 3142-3143, 2011.

ZHANG, W.; CHEN, J.; YANG, Y.; TANG, Y.; SHANG, J.; SHEN, B. A practical comparison of de novo genome assembly software tools for next-generation sequencing technologies. **Plos One**, v. 6, p. e17915, 2011.

ZHOU, Y.; CHOI, Y. L.; SUN, M.; YU, Z. Novel roles of Bacillus *thuringiensis* to control plant diseases. **Appl Microbiol Biotechnol**, v. 80, p. 563572, 2008.

ZHU, Y.; SHANG, H.; ZHU, Q.; JI, F.; WANG, P.; FU, J.; et al. Complete genome sequence of *Bacillus thuringiensis* serovar *finitimus* strain YBT-020. **J Bacteriol**, v. 193, p. 2379-2380, 2011.

ZIMMERMAN, S. B.; SCHWARTZ, C. D.; MONAGHAN, R. L.; PELAK, B. A.; WEISSBERGER, B.; GILFILLAN, E. C; et al. Difficidin and oxydifficidin: novel broad spectrum antibacterial antibiotics produced by *Bacillus subtilis*. I. Production, taxonomy and antibacterial activity. **J Antibiot (Tokyo)**, v. 40, p. 16771681, 1987.

ZWEERINK, M. M.; EDISON, A. Difficidin and oxydifficidin: novel broad spectrum antibacterial antibiotics produced by *Bacillus subtilis*. III. Mode of action of difficidin. **J Antibiot (Tokyo)**, v. 40, p. 1692-1697, 1987.

ZWICK, M. E.; JOSEPH, S. J.; DIDELOT, X.; CHEN, P. E.; BISHOP-LILLY, K. A.; STEWART, A. C.; et al. Genomic characterization of the *Bacillus cereus sensu lato* species: backdrop to the evolution of *Bacillus anthracis*. **Genome Res**, v. 22, p. 1512-1524, 2012.

MIX
Papier aus verantwortungsvollen Quellen
Paper from responsible sources
FSC® C105338

Printed by Books on Demand GmbH, Norderstedt / Germany